Thorkild Schiøler

ROMAN WATERWHEELS

Denne afhandling er af det filosofiske fakultetsråd ved Københavns universitet antaget til offentlig at forsvares for den filosofiske doktorgrad.

København, den 14. december 1971

BJARNE NØRRETRANDERS
decanus

Cover illustration: Investigation of Roman sāqiya at Tuna al Ǧabal in Egypt.

Printed by *Andelsbogtrykkeriet i Odense*, in Denmark. Clichés by *Fyns kliché Fabrik*. The translation from Danish has been made by *Pauline M. Katborg* and the graphic layout prepared by *Ib Jørgensen*. All the figures, except Figs. 1, 42, 43, and 49 have been drawn by the author.

ISBN 87 7492 090 1

Library of Congress

Catalog Card Number: 73-900 30

Acta Historica Scientiarum Naturalium et Medicinalium
Edidit Bibliotheca Universitatis Hauniensis Vol. 28

ROMAN AND ISLAMIC WATER-LIFTING WHEELS

by Thorkild Schiøler

ODENSE UNIVERSITY PRESS 1973

Drawings executed by the author on the basis of other pictorial material, not mentioned in the text.

Fig. 18a. J. Humlum: *Pirzada.* Copenhagen 1950. Fig. 27. A. Reifenberg: *The Struggle between the Desert and the Sown.* Jerusalem 1955. Fig. 28. The Unesco Courier. Fig. 56. E. H. Goetz: *Bilderatlas zur Kulturgeschichte Indiens in der Großmoghul-Zeit,* fig. 106, (detail from a miniature painting). Fig. 57. J. H. Marshall: *Excavations at Maṇḍor.* [Archaeological Survey of India, Annual Report 1909–1910, Calcutta 1914, plate 44a]. Fig. 58. Grace Dunham Guest: *Shiraz Painting in the Sixteenth Century,* Washington 1949, plate 44B. Fig. 59. Private photograph Klaus Ferdinand 1959. Figs. 69 and 87. The Library of the Academy of Art, Copenhagen. Fig. 35. Julio Caro Baroja: *Norias, azudas, aceñas.* [Revista dialectología y tradiciones populares, Vol. *X,* p. 42, fig. 4.]

Preface

During a holiday in Spain 20 years ago, I noticed some donkey-driven water-lifting machines used for irrigating the fields, and I realized that such machines were rarities and that their history might well reach far back in time.

I soon became fascinated by these contrivances and set myself the task of investigating their construction and origin without at first being aware of just how difficult this would be.

Over the next few years I had several opportunities of examining and surveying the most common types of water-lifting machines used in the Mediterranean countries, and at the same time I began a more or less systematic search for relevant literature in the libraries of Copenhagen.

Although the subject really belongs to the history of technology, most of the references are to be found in philological works. This has led me into fields far removed from the usual sphere of an engineer, so I sincerely hope that specialists will forgive possible shortcomings.

That my book does have a number of shortcomings is painfully clear to me. There should, for example, be a list of the Arabic terms applying to water-lifting machines, but after several abortive attempts at compiling one, I gave the task up as impossible. Many readers are likely to regret the lack of a detailed account of the development of water-lifting machines, but this would be encumbered with so much conjecture as to be practically worthless.

As far as concerns the history of water-lifting machines, I have concentrated mainly on Roman and Islamic sources, and this has meant that recent historical records have had to be dealt with more briefly.

Acknowledgments

I would like to thank the Danish State Research Foundation for two scholarships and another Danish foundation, which desires to remain anonymous, for a large grant, which enabled me to visit the Near East in my quest.

I am also deeply indebted to A. G. Drachmann, Ph. D., for his never-failing interest in my work and for invaluable advice and interpretations of Greek, Latin and Arabic texts.

Professor W. Müller-Wiener has my sincerest thanks for calling my attention to a number of very important Roman and Islamic ruins without which this work would have been incomplete.

Last, but not least, I owe a debt of gratitude to my friend, José M. Mañá, Director of the Archaeological Museum at Ibiza, for providing me with the inspiration for this work. Señor Mañá died in 1964, and this book is dedicated to his memory.

Copenhagen, 1973.

Contents

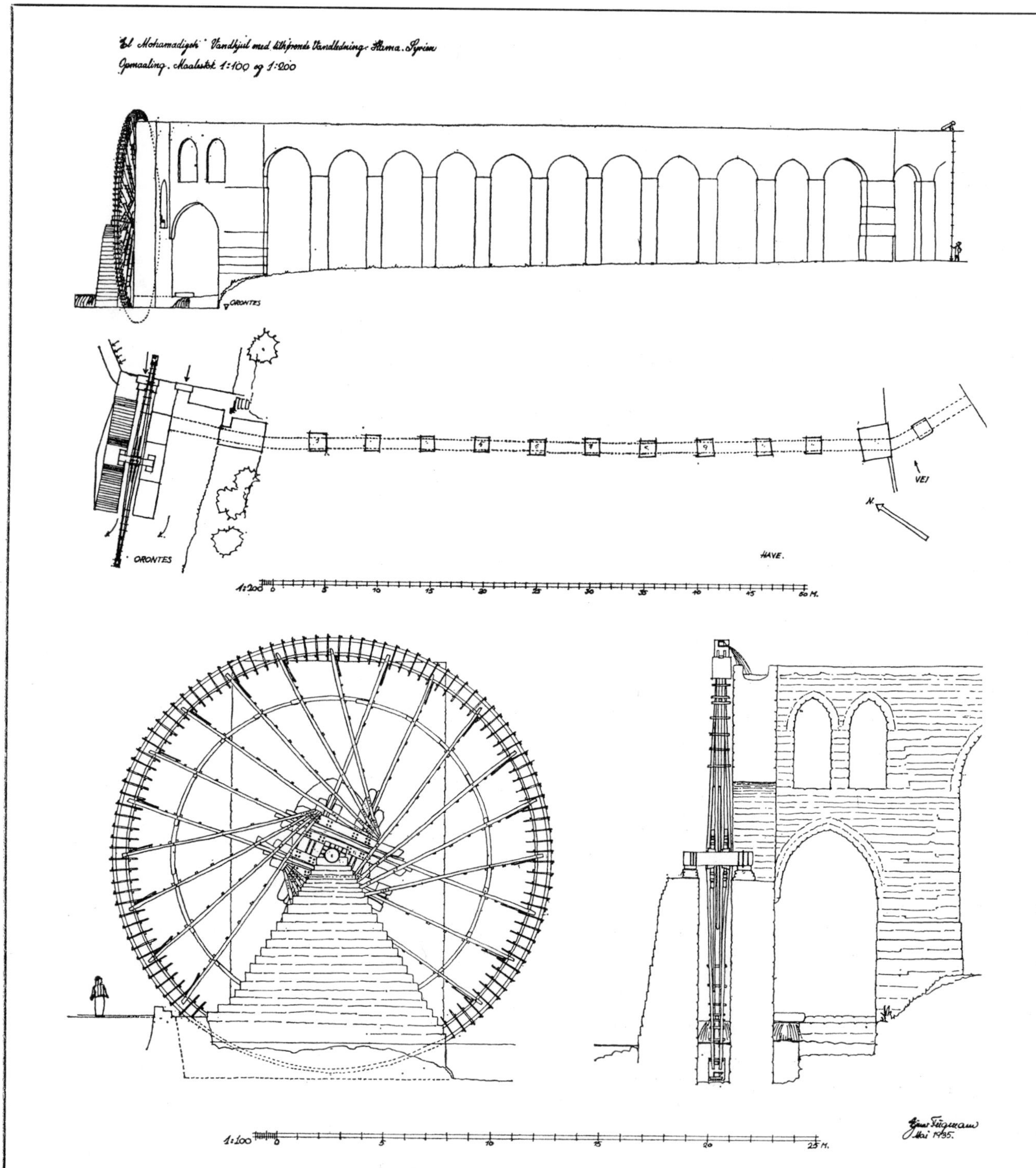

Fig. 1. One of the Great Wheels of Ḥamā in Syria. Scale drawing by Einar Fugmann 1935. The wheel is now (1969) in ruins.

Introduction

1. Definition of the subject

The subject of this book is water-lifting machines built of wood, clay, stone and rope, in which one wheel, or more, forms a principal component of the structure. Despite the narrowness of the field, the subject is both vast and complex, and other limitations have proved necessary. For example, the type of wheel to which "The Great Wheels of Ḥamā" belong is only touched upon in brief. See Fig. 1.

2. Two problems

Two main problems are encountered in a study of the literature: firstly, there are hardly any good technical drawings of the various types of structures, and secondly, a historical research is made difficult by the high degree of uncertainty regarding the origin of the machines. These problems are very closely related, since a solution of the former is necessary for solution of the latter. The fact is that the terminology used in the historical sources is largely incomprehensible to us, and only an intimate knowledge of the structures themselves can help us to interpret the texts.

3. Sources

In order to attain the necessary intimate knowledge of these machines, we have to study them in the field, and this provides the opportunity of contacting the local carpenter – the only man who really knows the machine and the relevant terminology.

Unfortunately, many of the old water-lifting machines have either disappeared altogether or have been replaced by modern pumps, and in such cases, our studies have to be based on old engravings or photographs. These are usually a poor substitute because the pictorial quality is often so poor that one can hardly even determine the type of machine depicted. A picture in which it is impossible to distinguish between a bolthead and a knot in the wood is worthless to a serious student of the subject. The engravings in earlier literature are very beautiful, but the artists can hardly be accused of having paid too much attention to precision, so these works have to be approached with great caution.

For the recent history of these machines our best sources are travel books and topographical works. Our knowledge of mediaeval machinery is usually derived from illuminated manuscripts and books that have to be regarded in the light of their author's purpose, which was to sketch new ideas rather than existing machinery. When we turn to Islamic manuscripts, the situation becomes even worse because the authors here were especially fascinated by the concept of perpetual motion and by the idea of a machine with a hidden mechanism. Fortunately, however, there are other sources that provide a more sober picture of the applications of such machines.

One extremely important source is the Greek papyri, which cover the period from *B.C.* 300 to *A.D.* 600. The greatest difficulty here is the interpretation of the terminology used, but a study of contemporary finds will doubtless help in the solution of this problem.

In the Roman and Islamic Empires there were a number of buildings where water-lifting machines formed part of the complex. From the few preserved for posterity it is possible to form a picture of the original structures from a study of the foundations.

The investigation of such ruins is greatly facilitated by bearing in mind the chapters in Vitruvius that deal with water-lifting contrivances, but it must be remembered that neither in Vitruvius nor elsewhere in the classical literature are there descriptions of a geared water-lifting machine.

Another extensive archaeological material is provided by the characteristic pots formerly used in paternoster wheels.

On the whole, our most valuable and reliable source of information is provided by classical archaeology. Thus, in Alexandria, there is a wonderful painting of a machine, which throws light not only on the history of water-lifting contrivances, but also on the history of technology itself.

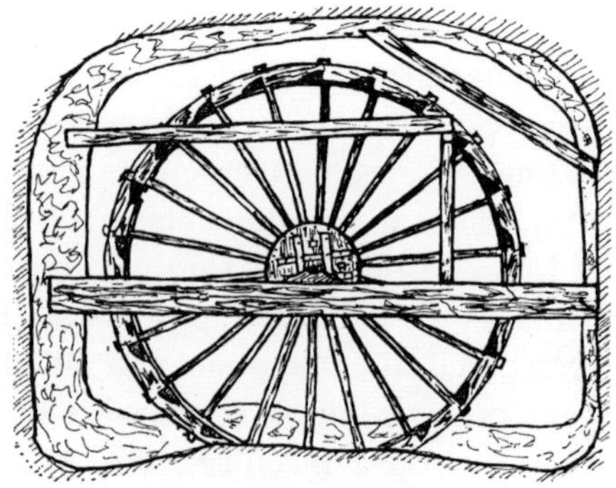

See p. 39, Fig. 35.

CHAPTER 1

Waterwheels: Classification and Technology

A review of past and present waterwheels in the Mediterranean world

To facilitate the student's approach to the subject it is necessary to have a classification, and this has the added advantage of solving the constantly recurring problem of terminology, since it permits the use of typological designations instead of names. On the other hand, it is difficult to remember twenty different labels, so to aid the memory, each designation includes the name of a geographical locality and a drawing.
The classification is based on the construction principle used. This is the system applied in Jorge Dias: *Aparelhos de elevar a água de rega.* Porto, Portugal 1953, p. 60, and, somewhat later, in Julio Caro Baroja: *Norias, azudas, aceñas.* [Revista de dialectología y tradiciones populares, Vol. *X* (Madrid 1954), pp. 29–160], and, in the next volume of the same periodical, Julio Caro Baroja: *Sobre la historia de la noria de tiro.* [Revista de dialectología y tradiciones populares, Vol. *XI* (Madrid 1955), pp. 15–79].

Waterwheels can be classified into two main groups:

I. Machines with a gear.
II. Machines without a gear.

Group *I* can be defined as machines in which the traction power of a draught animal is transmitted to a potgarland by means of a right-angle gear.

Group *II* may be defined as water-driven wheels or treadwheels, the rims of which are provided with a number of containers for lifting the water.

Group *I*

Group *I* can further be sub-divided according to the type of horizontal shaft:

A. Machines with a short shaft.
B. Machines with a long shaft.
C. Machines with an elevated shaft.

Fig. 2. Type IA, noria with short shaft.

These three types are shown in Figs. 2, 3 and 4. In principle, the cogs of a right-angle gear can be arranged in two ways, viz:

1. Radially projecting cogs.
2. Cogs arranged parallel to the shaft.

We thus have six different types.
Turning to the paternoster chain, we find two different systems:

a. Earthernware pots or tins, firmly lashed to two endless ropes.
b. Rectangular buckets fastened to a ladder-like iron chain.

This last sub-division increases the number of types from six to twelve.

Fig. 3. Type IB, sāqiya with long shaft.

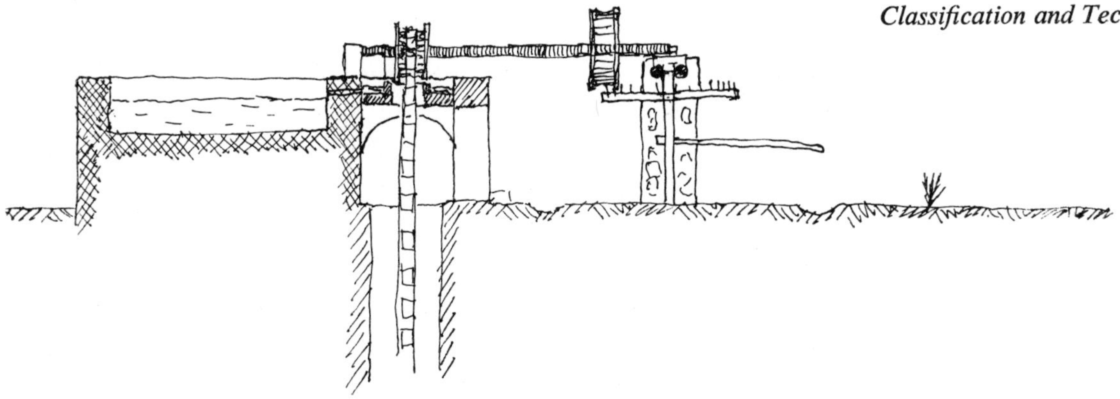

Fig. 4. Type IC, sāqiya with long, elevated shaft.

At the beginning of this chapter, the geared machines were given a rather narrow definition, which means that the classification is representative. If we were to broaden the definition, the classification would become exhaustive, but then not representative. It has therefore been found best to abide by the definition given and to describe variations simply as "other geared machines". Only two types of machines are included in this category:

1. Geared machines powered by a draught animal, but with the paternoster chain replaced by a wheel with a compartmented rim. See Fig. 29.
2. Geared machines powered by a draught animal, but with the paternoster chain replaced by a paddle-wheel, the blades of which follow a spiral course. See Figs. 54a and 54b.

We thus have a total of 14 different types of machines in group *I*. The first 12 of these are arranged systematically, while the last two do not quite fit in, which is unfortunate since they are both very common in Egypt.

Group *II*

Group *II*, which is only of secondary interest here, comprises water-wheels without a gear. The definition of this class of wheels on page 11 includes the basic sub-division, the following designations being used:

A. Water-driven wheel.
B. Treadwheel.

The term water-driven wheel indicates a paddle-wheel turned by the current of a river. See Figs. 1 and 32. A treadwheel is a large wheel turned by the tread of one or two labourers. See Fig. 100.

Two types of containers are used:

1. Slender pots or hollow bamboo canes lashed to the rim of the wheel.
2. Partly closed compartments forming the actual rim of the wheel. See Fig. 34. In this case, the structure is known as a compartmented wheel.

Two types of machines remain, which do not fit into the classification, but which are a hybrid of the two groups. These we will call "other machines":

1. A treadwheel and a potgarland sharing the same shaft. See Fig. 95.
2. A composite hand- and treadwheel designed to carry a potgarland. See Figs. 56 and 57.

A map at the end of this book shows the geographical distribution of the two main groups. See Fig. 114.

Vignette No. 1. This pulley device lies outside the scope of this book.

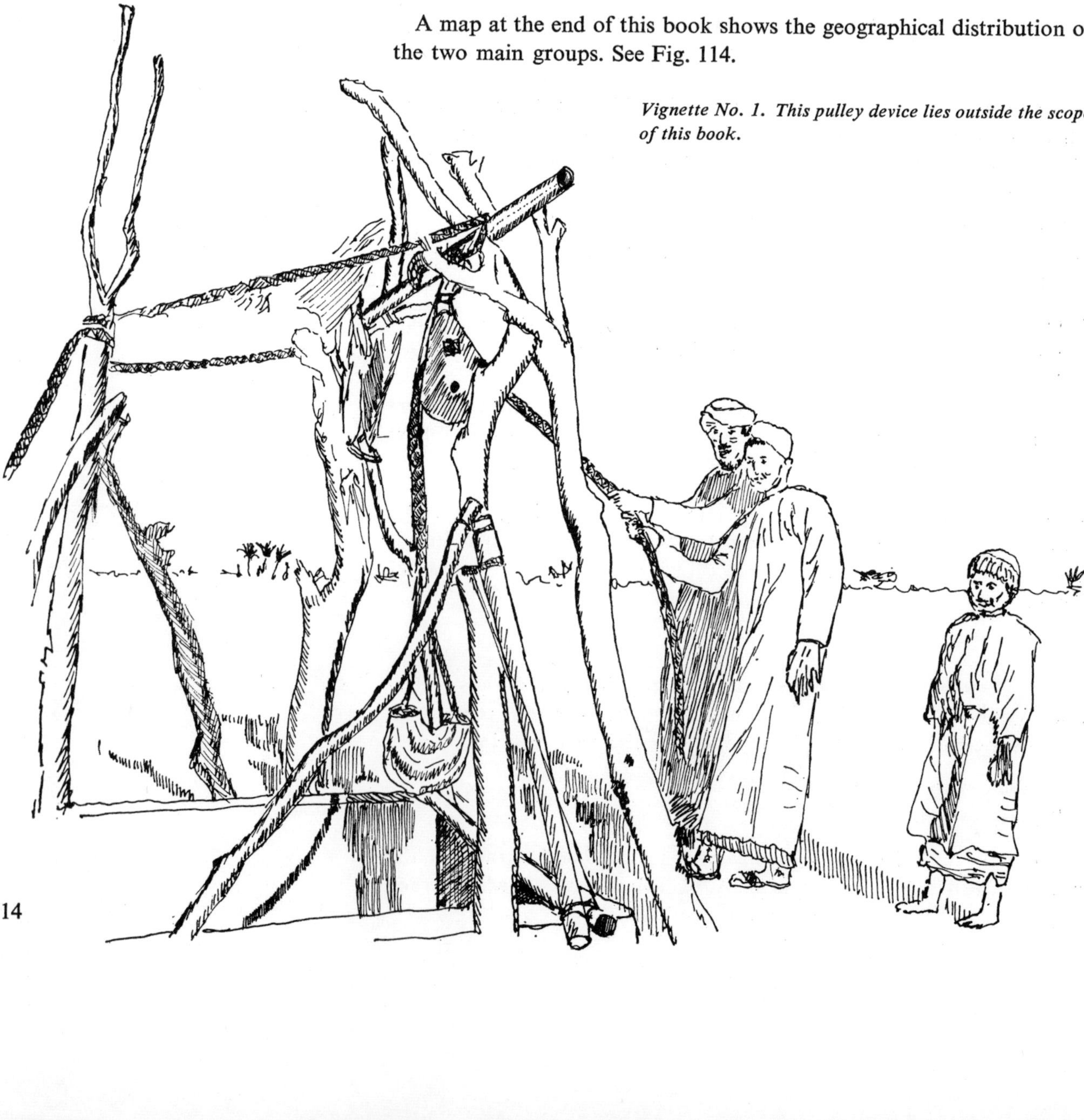

Classification of Waterwheels

I Machines with gear

A. Short shaft (noria)

1 Parallel cogs	*a* Potgarland	Ibiza
	b Bucket chain	Maʿarrat
2 Radial cogs	*a* Potgarland	Salḥin
	b Bucket chain	Aleppo

B. Long shaft (sāqiya)

1 Parallel cogs	*a* Potgarland	Hala
	b Bucket chain	Ḥaswa
2 Radial cogs	*a* Potgarland	Miʿtimdiyya
	b Bucket chain	Fayyūm

C. Elevated shaft (sāqiya)

1 Parallel cogs	*a* Potgarland	Formia
	b Bucket chain	Palestine
2 Radial cogs	*a* Potgarland	India
	b Bucket chain	Syria

II Machines without gear

A. Water–driven wheel (hydraulic noria)

1 Pots	Hīt
2 Compartmented rim	Ḥamā

B. Treadwheel

1 Pots	Ríotinto
2 Compartmented rim	Ostia

We will now discuss the individual types of machines. In order to avoid repetition only the noria-type from Ibiza and the sāqiya-type are dealt with in detail.

Group *I*. Machines with gear

Short shaft

Type *IA1a*
Ibiza

This type of noria is native to Spain, Portugal and Morocco, and is characterized by its short shaft. See Figs. 5, 6, 7, 55, 69, and 113.

La Mancha

South of Madrid, norias were still numerous in 1960, and in the famous province of La Mancha, as many as 19,000 were counted in 1918. Otto Jessen: *La Mancha.* [Mitteilungen der Geographischen Gesellschaft in Hamburg, Vol. 41 (Hamburg 1930), pp. 180–184]. In about 1900 there were no less than 4,000 norias in operation on Majorca, but only a few of these remain today (1963). As far as Portugal is concerned, this subject has been dealt with in detail by Jorge Dias (cf. p. 11), and it appears that almost all types of water-lifting machines are represented there. It is a pity that similar books are not available for the other Mediterranean countries.

Ibiza

In the neighbourhood of the town of Ibiza there are many norias (1955), sometimes only a hundred metres or so apart, corresponding to one for each farm. In these norias, both the machine and its water tank are raised one or two metres above ground level, which means that the tank also acts as an elevated cistern from which water can be led all over the farm.

Fig. 5. Noria type IA1a, Ibiza. The lantern pinion engages with the paternoster wheel. The pots empty the water into the trough standing inside the wheel.

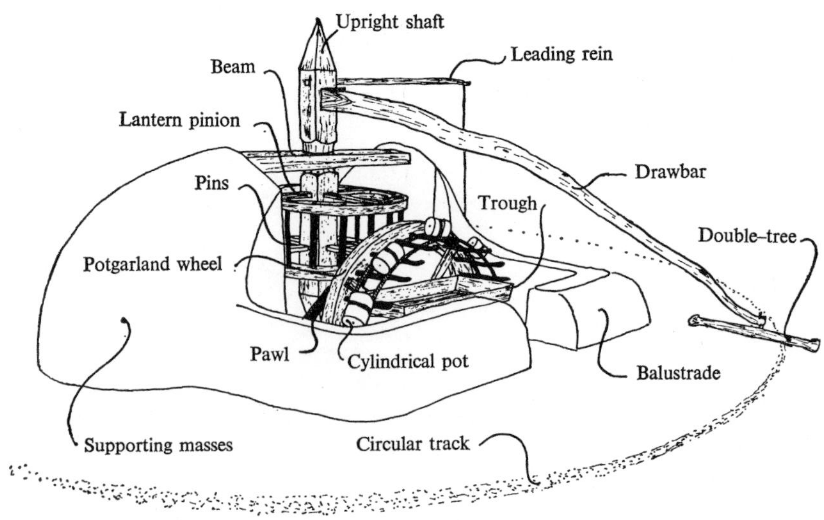

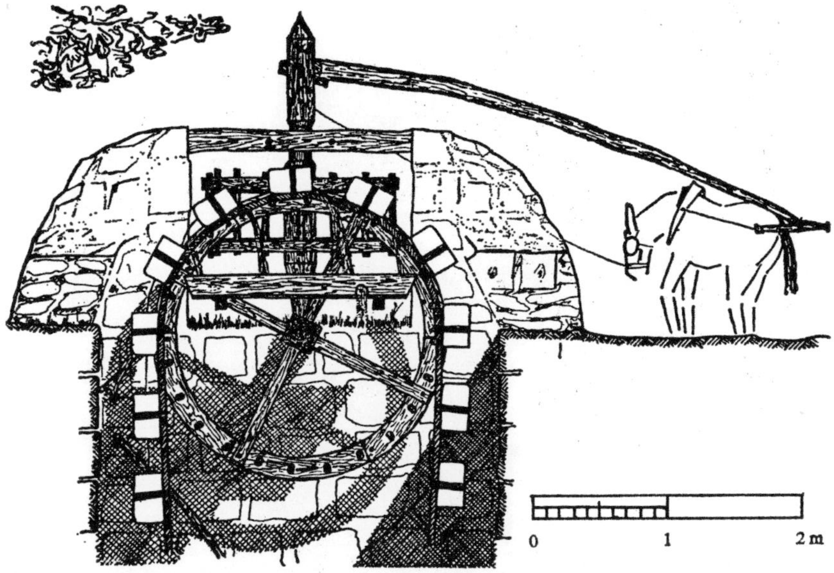

Fig. 6. Noria Ibiza, Balearic Islands. Scale drawing by the author, 1955. From the wooden trough the water runs through a pipe leading under the circular mule-track to a large water tank.

The total area of a farm seldom exceeds one hectare and of this the tank occupies about 100 square metres and the machine a further 100 square metres. The well, which is rectangular, has a depth of approximately 4 metres and is lined with stone. The edge of the well is provided with a balustrade, which curves elegantly into the two supporting masses. The building material used is rough-hewn volcanic rock, a smooth surface being achieved by means of mortar and plaster. In June of each year, the womenfolk whitewash the masonry. The water tank is built in the same way, except that the interior surface is given a cement rendering. The internal dimensions of the water tank are 10 × 10 × 1 metres, giving a capacity of 100 cubic metres or 1 cm-ha, a unit preferred by irrigation engineers. At one corner of the water tank, steps lead down to the water, and here the women wash the newly shorn wool. Drinking water is taken directly from the machine. The water raised by the wheel flows into the water tank through a pipe laid under the circular mule-track. Every evening, the tank is opened and the water is led out to the fields through a large number of walled channels.

In Salvator's two-volume work on the Balearic Islands this scene is depicted in some very fine engravings, which are also remarkable for their exactitude. Ludwig Salvator: *Die Balearen, geschildert in Wort und Bild.* Würzburg und Leipzig 1897, Vol. *I*, p. 7 and p. 229.

Figs. 6 and 7 show a scale drawing of a noria which was still in use in 1955, but which I found in ruins just a few years later.

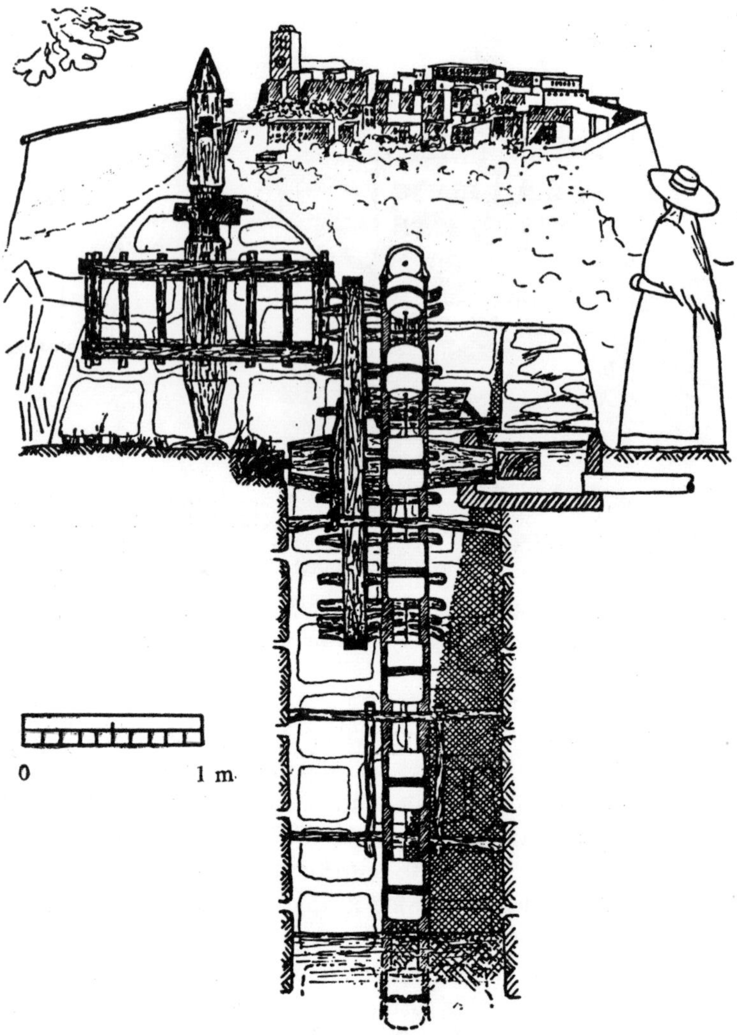

Fig. 7. Noria Ibiza. Scale drawing by the author. The same machine as in Fig. 6, but seen from the other side.

The crux of these machines is the gear, which has a single function – that of altering the motion from horizontal to vertical. The gear itself consists of a lantern pinion and a large cog-wheel. In the following, the latter will be referred to as the potgarland wheel, since it also operates the paternoster chain.

The most convenient way of describing such a machine is to follow the power transfer through the machine, starting with the draught animal and ending where the water leaves the machine. The draught animal is usually a mule or a donkey wearing on its shoulders and neck a collar harness that transmits the power through two traces to a double-tree fastened to the drawbar. The drawbar passes through a hole in the upright shaft. The leading rein, which is attached to a second bar protruding from the shaft, forces the animal to follow the circular track, although it cannot see this because its eyes are blinkered.

The lantern pinion is fixed to the upright shaft by means of two sets of spokes, one for the upper rim and one for the lower. The spokes are made in one piece from bars equal in length to the outer diameter of the rims and countersunk inside the shaft so that they bisect each other in the same plane. To achieve this, the hole through which one bar passes through the shaft is made twice the height of the hole for the other bar. When the spokes are in place, the larger hole is wedged to keep the spokes locked in position.

Both wheel-rims of the lantern pinion are made by the carpenter from six or seven pieces of wood, cut to a template. The pieces are joined together by double-lipped and notched table scarves. See Fig. 8. The scarf joint method of assembly is efficient as long as the wood is sound but is valueless when it dries out. It is therefore quite common to see peasants stiffening such joints by means of wooden scarves. The pins of the lantern pinion are slightly conical in shape and are of a soft wood because they are easy to make and the peasant puts in new pins approximately every other year. The cogs with which the pins mesh are difficult to make so a hard wood is used that will last for many years.

Fig. 8. Double-lipped and notched table scarf. An iron wedge keeps the two parts of the rim firmly together.

Before we leave the lantern pinion there are a few other details concerning the upright shaft that are of interest. At its base, the shaft ends in an iron gudgeon, which is bedded in a stone. The end of the shaft is provided with an iron ring to prevent it from splitting. See Fig. 9. The

Fig. 9. Lantern pinion, with pins, rim with spokes, and upright shaft with iron ring and gudgeon.

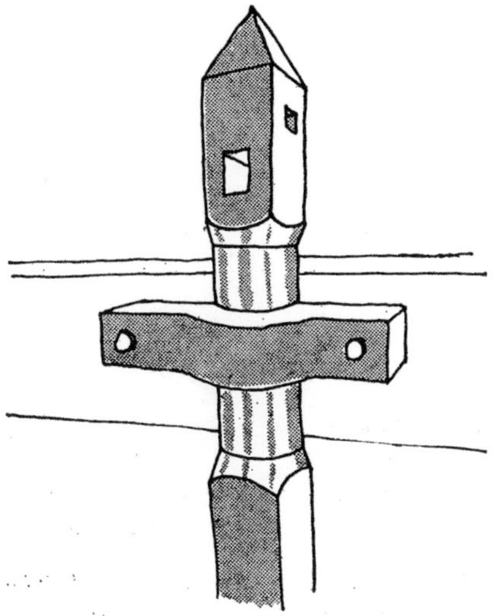

Fig. 10. The top bearing consists of a horizontal beam and a bearing lining.

Fig. 11. Paternoster wheel with pawl and potgarland. Majorca.

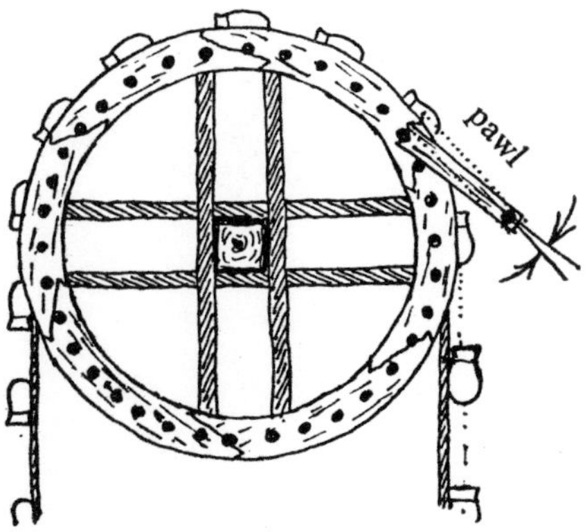

upper bearing for the shaft is located about 20 centimetres above the lantern pinion and consists of a beam with a yoke or liner fixed in position by two wooden spikes. The arrangement is supported on the two characteristic supporting masses mentioned earlier. See Fig. 10.

The potgarland wheel has two functions: it is primarily a cogwheel, but it also forms the drum that carries the potgarland. Each cog and carrying peg are made in one, as shown in Fig. 7. The potgarland wheel of the Spanish noria has only one rim, which is made from six pieces, and four spokes made from two continuous bars equal in length to the diameter of the wheel. The short shaft is very thick and is provided with a gudgeon at each end, turning in wooden blocks.

In order to prevent the wheel from going into reverse, the machine is provided with a pawl mechanism, which acts on the cogs of the potgarland wheel. See Fig. 11. To appreciate the vital function of this pawl it is only necessary to remember that the draught animal is subjected to a constant pull both when moving and when standing still. This pull is exerted by the part of the potgarland carrying the full pots. The pawl is activated in two cases – when the animal is to be unharnessed and in the event of the harness or traces breaking. Without the pawl the machine would turn backwards at great speed and, after one revolution, the drawbar would hit the draught animal on the head. At the same time, many of the pins of the lantern pinion would break and the pots smash. A local carpenter once told me that several conscientious donkeys had been killed because of defective pawls. However, although most of the norias I saw in Ibiza were more or less defective, their pawls were always in order. In this connexion, it should be noted that not all waterwheels are provided with a pawl mechanism. In places where the lift is less than 2 metres, there is practically no risk of a serious accident because the drawbar would move too slowly to kill the donkey. The draught animals are completely familiar with the operation of the pawl. If we watch one, we will see it turn its ears in the direction of the tick-tock of the pawl when it wants to stop. When it hears the sound the animal halts and takes a step backwards, at which point the pawl mechanism takes over the pull exerted by the full pots. If the animal did not await the sound of the pawl, it would have to move back 1.5 metres, which is against its nature. Now let us turn to the potgarland, which consists of two endless ropes spaced according to the diameter of the cylindrical pots. Besides serving as pot-carrier, the ropes act as a buffer between the pots and the carrier-pegs of the wheel, *i.e.* the pots are carried on the ropes, not between them. If the pots were carried in the latter manner, the edge of a pot might catch on a carrier-peg and break.

The pots have a small vent at the bottom to permit air to escape when the pot hits the water bottom up. Unfortunately, this vent gradually becomes clogged with leaves, and when that happens the pot floats on the water, the next pot hits it and they both break. The vent appears to be a disadvantage because some of the water runs out of it. However, there is really very little loss because most of the water just runs down into the next pot. Rather more water is lost as the pots discharge their contents at the top of the wheel, because some of it hits the carrier-pegs and splashes down outside the trough.

A discussion of the *modus operandi* of the potgarland must include a description of the two guide-rods which keep the potgarland on the right course up to the potgarland wheel. These two rods are placed on either side of the potgarland and at an angle of about 45° to it. Without the rods, the potgarland would slide out along the carrier-pegs and finally fall off them. The constant bumping of the pots against the guide-rods results in deep wear marks on either side of the pots (cf. p. 102).

"Inside" the potgarland wheel there is a large wooden trough supported by two brackets embedded in the balustrade of the well.

Efficiency

The efficiency of a machine is defined as the ratio between the output power and the input power. This ratio gives only the efficiency of the machine itself; to obtain an idea of the total efficiency it would also be necessary to take into account the efficiency of the draught animal. However, for the sake of simplicity, the efficiency of the machine alone will be considered in the following.

In the case investigated, the tractive force exerted on the drawbar varied between 50 and 250 newtons, *i.e.* between 5 and 25 kilograms force. The mean value was found to be 120 newtons. The large variations were caused by the fact that three or four pots were missing from the potgarland.

Calculation

The input power is now determined as the force multiplied by the velocity or, to be correct, as the torque multiplied by the angular velocity. As the force is not exerted at right-angles to the drawbar, we have to multiply the tractive force by the cosine of the angle between the traces and the drawbar:

$$P = T \times \omega = F \times v \cos\alpha$$

$$P = F\frac{2\pi n}{60} R \cos\alpha = 120 \frac{2 \times \pi \times 1.7}{60} 3.65 \times 0.95 = 75 \text{ watts}$$

where $n = 1.7$ rpm, $\cos\alpha = 0.95$, and the radius of the circular mule track $R = 3.65$ m; the remaining notation is as follows:

P = power, F = force, v = velocity, T = torque, ω = angular velocity.

For the output power we have to measure the rise in water level in the water tank within a certain period and determine the water-head. The flow was found to be 67 l/min, corresponding to 670 newtons/min, and the head was estimated to be 4 m. The output power is then calculated as the flow multiplied by the head: $670 \times 4/60 = 44$ watts.

The efficiency of the machine is then:

$$\eta = \frac{44}{75} = 0.6$$

This value is significant in that it bears comparison with the efficiency of modern pumps. As a rule of thumb, the flow from a defective, mule-drawn noria will be 4 m^3/h for a head of water of 4 m. If the head is doubled, the flow must be halved.

As we watch the potgarland and its wheel, the following question arises: why does the potgarland not slip on the wheel when the pull exerted by the full pots greatly exceeds that exerted by the empty pots? We shall find the answer by investigating whether the following inequality is fulfilled:

$$\frac{S_1}{S_2} < e^{\mu\alpha}$$

where S_1 is the pull exerted by the full pots, and S_2 is that exerted by the empty pots. According to the engineering manuals, the coefficient of friction is of the order of magnitude of 0.4 to 0.5. The potgarland circumscribes an angle α of slightly more than π or 180°. The value e is the base of natural logarithms. By substituting on the right-hand side of the inequality, we obtain values from 3.5 to 5. In the noria discussed above, the ratio between the pull exerted in the two directions was $800/300 = 2.7$, *i.e.*, less than the permissible value, which means that the rope does not slip. On the other hand, while I was in Egypt I noticed a potgarland in which the pots had been replaced by buckets, and here the rope did tend to slip. The ratio in this case between the pull in one direction and that in the other was $350/100 = 3.5$, *i.e.* in the vicinity of the permissible value.

Type *IA1b*
Maʿarrat al Nuʿmān

This type, which is found in Syria (1969), is very similar to the Spanish noria, except that a bucket chain and bucket wheel are used.

On the road from Ḥamā to Aleppo there is a small town called Maʿarrat al Nuʿmān, which once had a large number of norias. Unfortunately, they are now no longer in use and are therefore in ruins, but I did manage to find one in a sufficiently good state of repair to allow me to make a scale drawing of it. See Figs. 12 and 13.

The bucket wheel consists of 10 compartments, each with an outlet. The wheel itself is held together by two hoops and a large number of long bolts and fittings. Far less water is wasted with this construction than with the former. In the village of Tall Minnis, 5 kilometres east of the town, I saw this type of wheel in operation. It stood in the middle of the market place, raising drinking water for the villagers.

Fig. 12. Defective noria at Maʿarrat al Nuʿmān; on the road from Ḥamā to Aleppo [Ḥalab]. The wheel is provided with open compartments into which the buckets empty their water. The bucket chain has gone, and the front "hoop" has fallen off. The stone on the right is what remains of a Roman column.

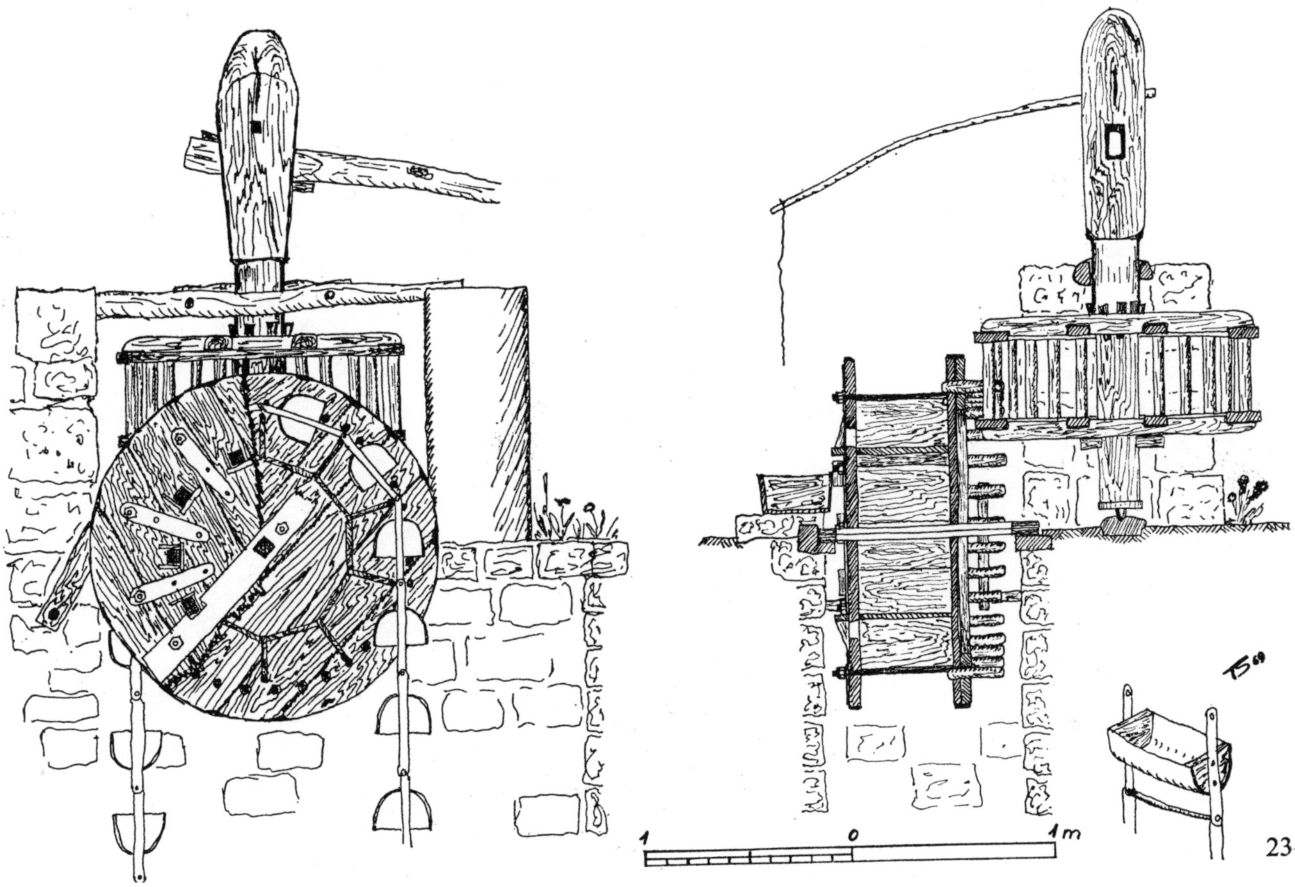

Fig. 13. Reconstruction of the defective noria on the road from Ḥamā to Aleppo. The left-hand drawing shows a partial section of the bucket-chain wheel, and the right-hand drawing shows a section of the bucket-chain wheel and the lantern pinion.

About 4 kilometres south of the Citadel of Aleppo lies a district called Salḥin, where there were still many norias of this type when I visited the area in 1969. See Fig. 14. The brickwork surrounding the well and the water tank is built along the same lines as in the norias of Ibiza, but the wooden machinery differs considerably from the Spanish version.

Type *IA2a*
Salḥin

The upright shaft, which is very long and very heavy, rotates in a bedding stone. The cogwheel itself is a simple construction, consisting of four heavy beams fixed in position by means of wedges and diagonals. See Fig. 15. The upper bearing consists of two long beams and two small crossbars, which together form a square gap through which the shaft passes. Two heavy blocks of stone are used to weigh down the beams and keep them in position. See Fig. 16.

Fig. 15. The upright shaft and cog-wheel. The bedding stone at the bottom must be subjected to a load of more than half a ton.

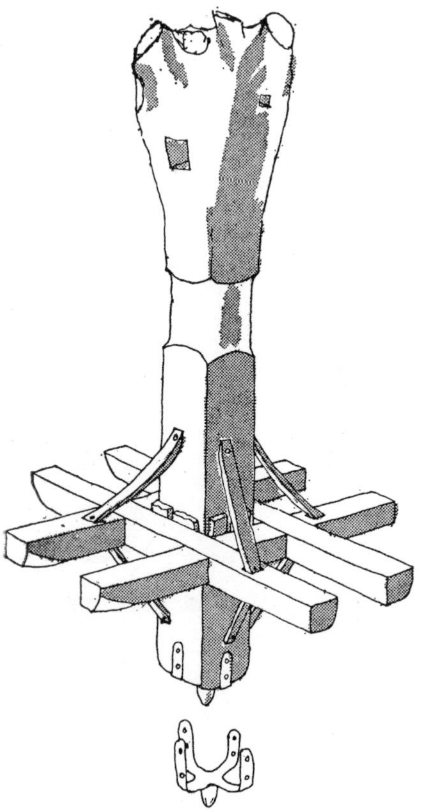

Fig. 14. Noria at Salḥin outside Aleppo. The upright shaft has a height of almost 3 metres.

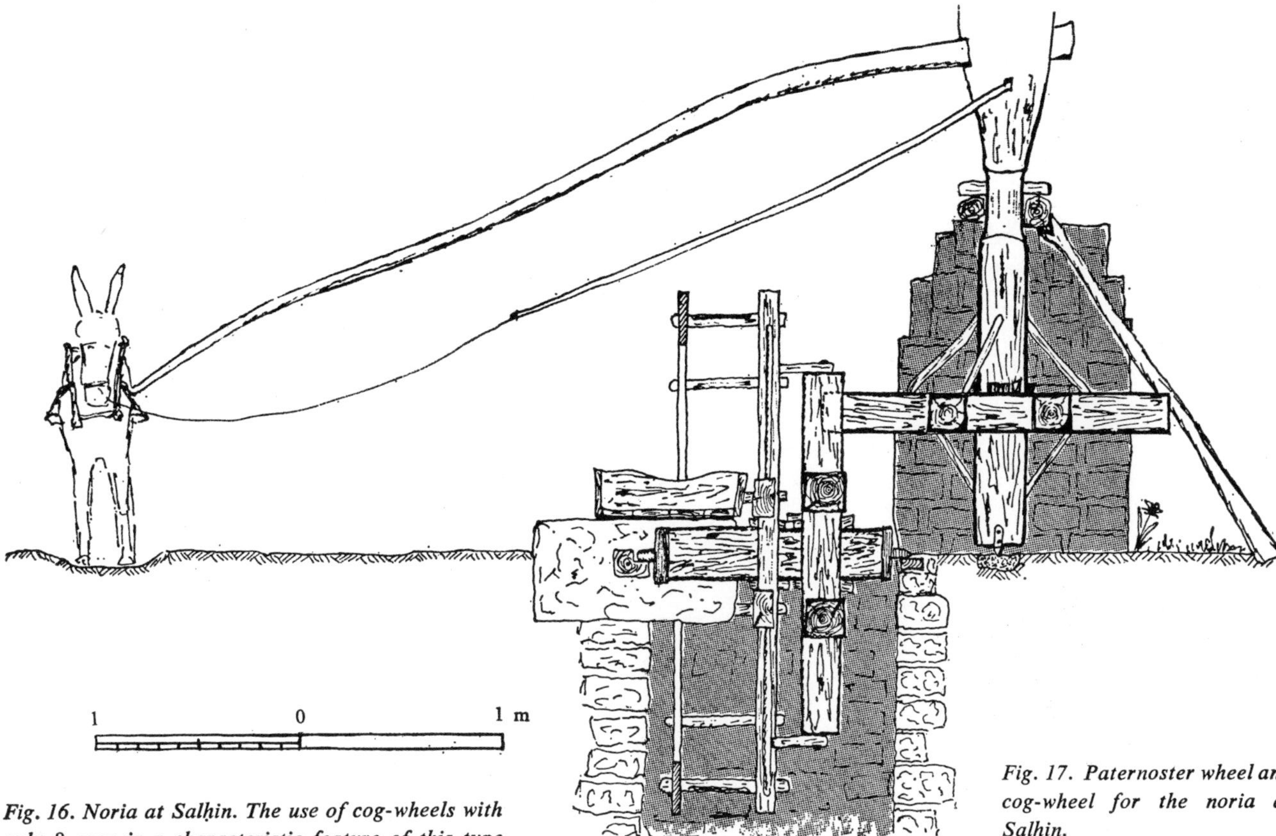

Fig. 16. Noria at Salḥin. The use of cog-wheels with only 8 cogs is a characteristic feature of this type of waterwheel. Forty years ago, potgarlands were still in use, but these have now been replaced by chains and tin buckets. These have been omitted from the drawing.

Fig. 17. Paternoster wheel and cog-wheel for the noria at Salḥin.

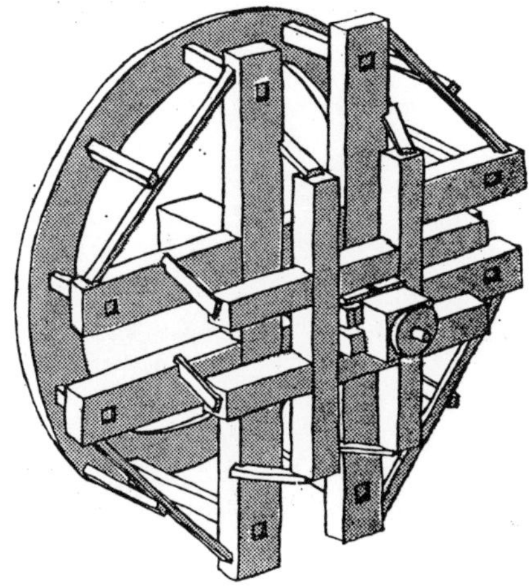

The horizontal shaft carries two wheels, see Fig. 17, one of which acts as cogwheel, and the other as potgarland wheel. The latter will be described later as it is identical with the drum in the Egyptian sāqiya, type *IB2a, Al Miʿtimdiyya.*

The potgarland originally consisted of hemp ropes and earthenware pots, but nowadays iron chains and buckets are used instead.

This noria had no pawl because there were no cogs on which it could play, and no guide-rods were used. All in all, it was very primitive and had to be supervised by two women.

The machine was in operation while I measured it – a rather hazardous occupation – but I made good use of the opportunity offered by stoppages caused by a couple of buckets getting caught up in the potgarland wheel.

Type *IA2b*
Aleppo

This type has the same gear construction as the previous type, but the potgarland wheel is replaced by a bucket wheel and bucket chain. According to Moussly, this machine is called a ḥallāf. Nazim Moussly: *Le problème de l'eau en Syrie.* Lyon 1951, p. 144.

Fig. 18a. Sāqiya with long shaft. Note the lantern gear construction. West Pakistan.

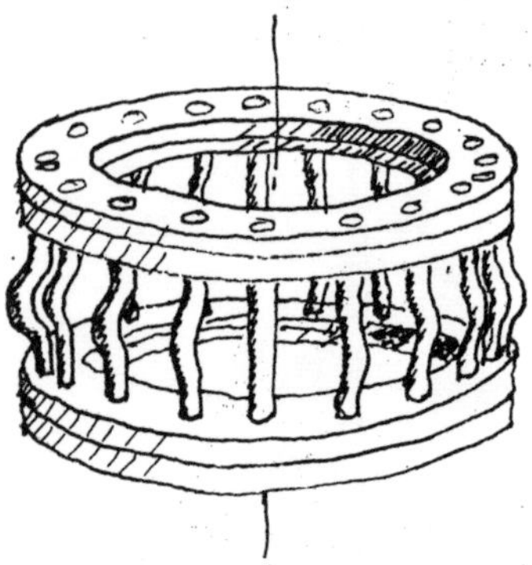

Fig. 18b. The lantern pinion.

Fig. 19. Iron sāqiya. Iraq, 1969.

Long horizontal shaft

We now come to the types with a long horizontal shaft. Owing to the method of classification adopted, we have to deal with a rare type of wheel before the Egyptian sāqiya.

The fascinating point about this machine is its use of a right-angle gear, which represents an advanced level of joinery. See Figs. 18a and 18b. The rims are constructed of laminated wood and the pegs are curved, both details that make the construction rigid and compact. Type *IB1a, Hala.*

As norias are still being built in many countries, attention should be drawn to this joinery method and, in the case of wells with a depth of more than 4 metres, the Spanish design should be borne in mind. The fact that many countries prefer norias made of iron is undoubtedly due to the poor quality of the joinery in earlier norias. In the next example we shall examine a machine made of iron. This is type *IB1b, Ḥaswa.*

Although machines made of iron really lie outside the scope of this book, there is good reason to include this example because it is a true copy of its predecessor, which was made of wood. The drawing in Fig. 19 has been made on the basis of my survey of a dawlāb (sāqiya) near Ḥaswa, between Baghdad and Babylon. In this region there are hundreds of machines, but they are all made of iron; for the wooden version we have to go to Northern Iraq. There is a photograph of such a machine in an illustrated work dating from 1968. According to the caption, the picture is from Palestine, but this must be an error because there is no

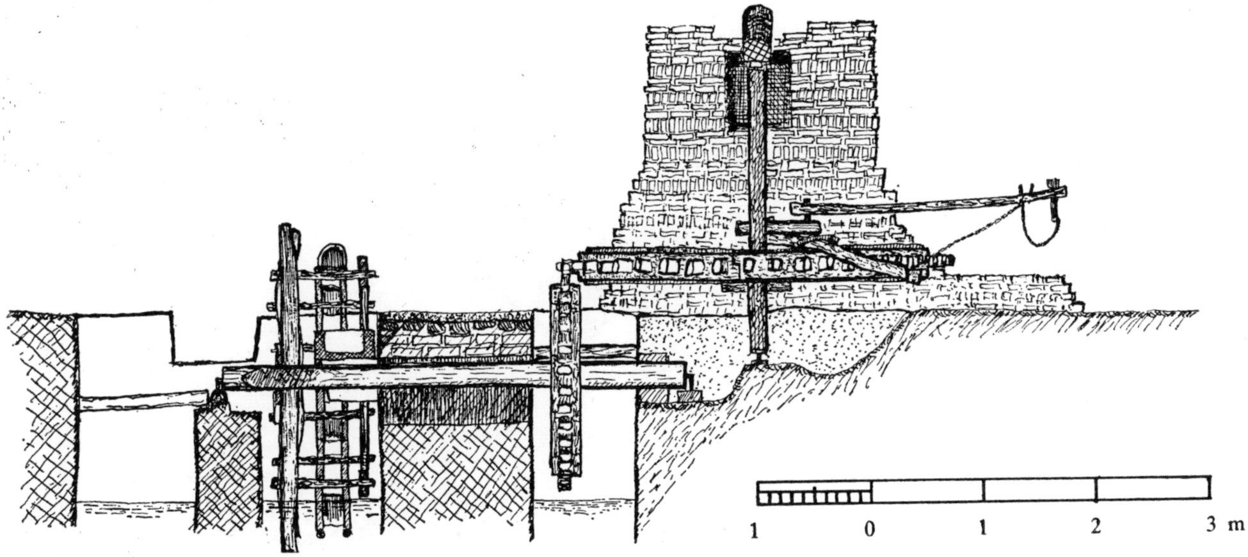

Fig. 20. Cross-section of the sāqiya at Miᶜtimdiyya near Cairo. Drawn by the author in 1964. For details see Figs. 21, 22, 23 and 24.

doubt at all that it actually comes from Northern Iraq. D. et J. Sourdel: *La civilisation de l'Islam classique*. Paris 1968, fig. 96. It is interesting to note that this construction uses a lantern pinion with one wheel rim instead of two. See also Vignette No. 2 on p. 41.

Type *IB2a*
Al Miᶜtimdiyya

The best known of all water-lifting machines is the Egyptian sāqiya. The monotonous creaking of the cog-wheels can be heard all the way from the shores of the Mediterranean to deep in the Sudan, where:

> In 1943, the number of waterwheels in the Northern Province was 9,456 ... the figure has declined less than might have been expected.

K. M. Barbour: *The Republic of the Sudan*, London 1961, p. 141 and fig. 41.

Let us now return to Egypt, to the town of Al Miᶜtimdiyya, which lies outside Cairo, about 5 kilometres west of Cairo Tower, for here there was an old sāqiya of great interest. See Fig. 20. The well, which is only 2 metres deep, is divided into two sections, so there is actually room for two sāqiyas. See Fig. 21. Not far away lie the ruins of another well, in which three sāqiyas once worked. In both cases, the wells are made of burnt red brick, while the surrounding walls consist of mud bricks.

Turning now to the machine itself, we shall follow the power input, from the ox, through the gear, to the potgarland. The sāqiya in question is drawn by an ox moving extremely slowly and bearing on its shoulders a neck-bar, the other end of which is pinned to a small beam cantilevered

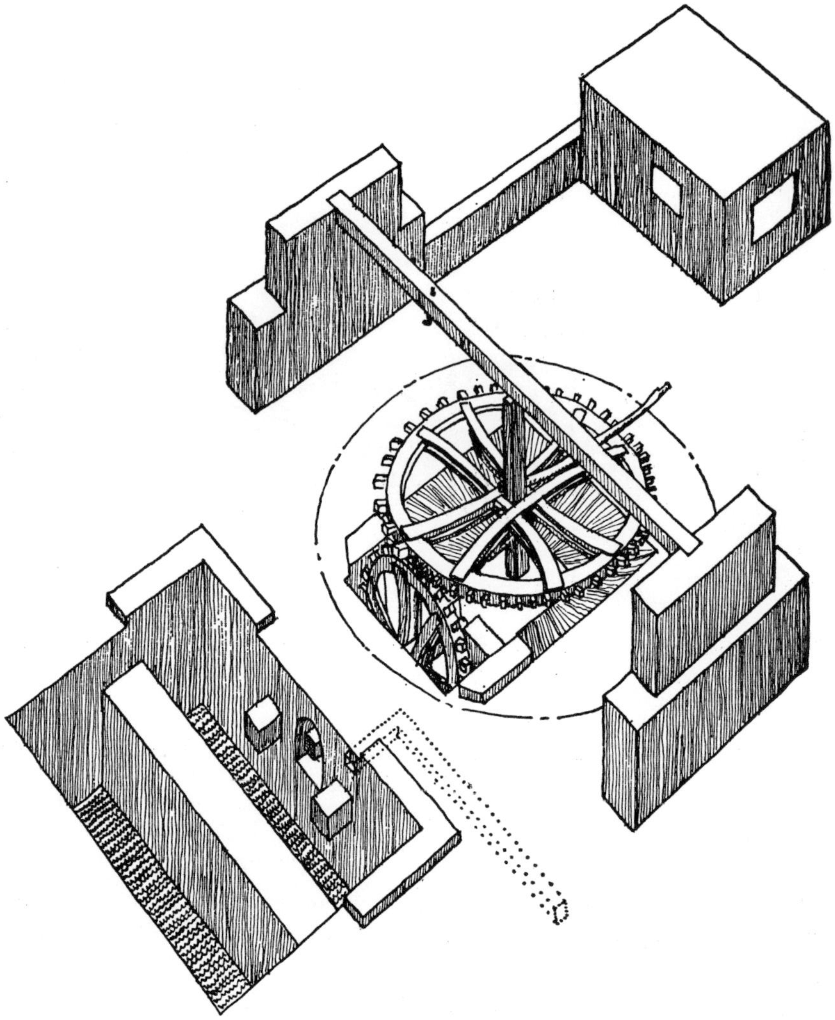

Fig. 21. Perspective of the sāqiya. The bucket-chain wheel has been removed. Miʿtimdiyya.

out from the upright shaft. See Fig. 22. If the sāqiya were drawn by two oxen, the neck-bar would be replaced by a yoke. See Fig. 109.
In order to keep the ox on the circular track, the outer end of the neck-bar is provided with a wedge and rope arrangement, forming a halter. The pull of the ox is transmitted by a chain to the drawbar, which is fixed to the cogwheel a quarter turn behind the neck-bar. As stated earlier, the cogwheel is in turn fixed to the upright shaft, the upper end of which has a gudgeon turning in a long palm beam. The lower end rotates in a bedding stone or wooden block, as shown in Fig. 88. The rim of the cogwheel is fixed between an upper and a lower set of spokes. The rim itself consists of six pieces, joined together in a manner that prevents radial displacement. Forty-eight holes are then made in the rim, corresponding

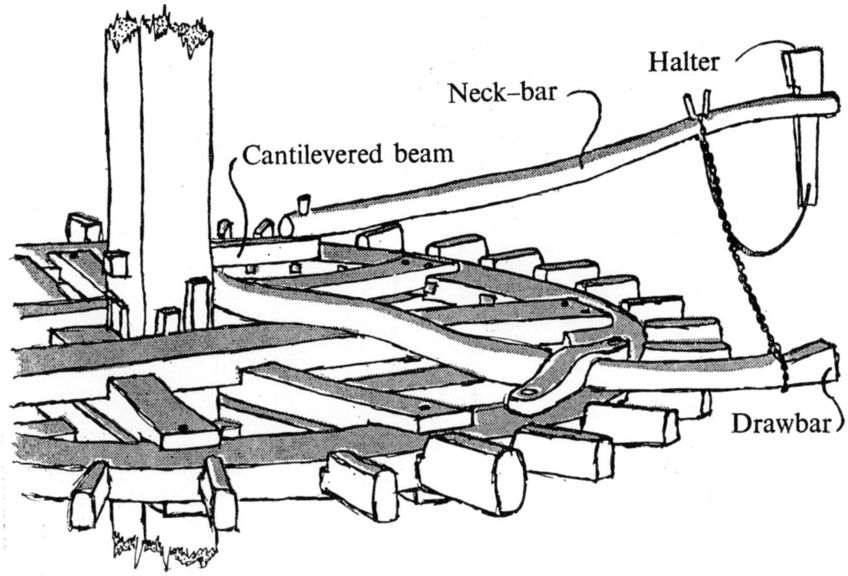

Fig. 22. The large cog-wheel, including the long neck-bar with the chain that transmits the pull to the cog-wheel.

to the number of cogs. Some of the holes coincide with the rim joints, which are therefore spoiled. However, the real assembly mechanism for the rim is provided by the spokes, which act as scarves. The entire weight of the the gear-wheel is supported by a 60 centimetre long transverse rod, and the wheel is centred by means of wedges between the spokes and the upright shaft.

Fig. 23. The horizontal shaft with small cog-wheel and potgarland wheel.

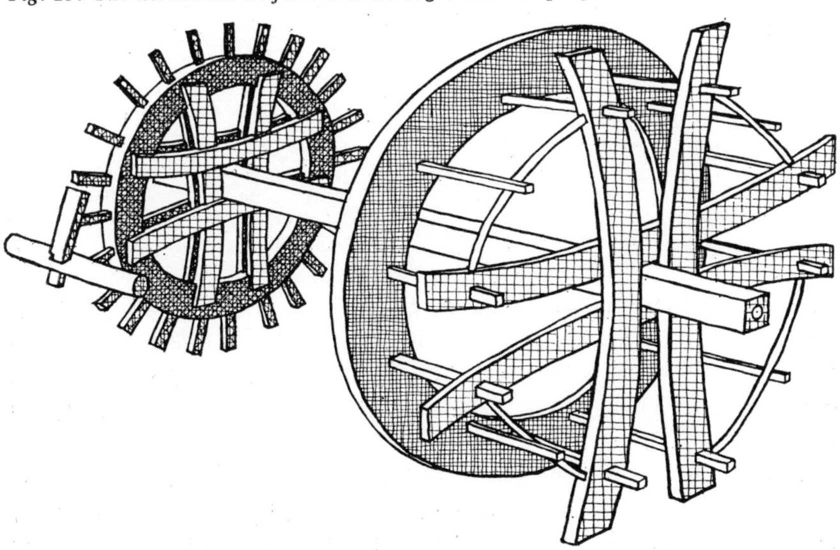

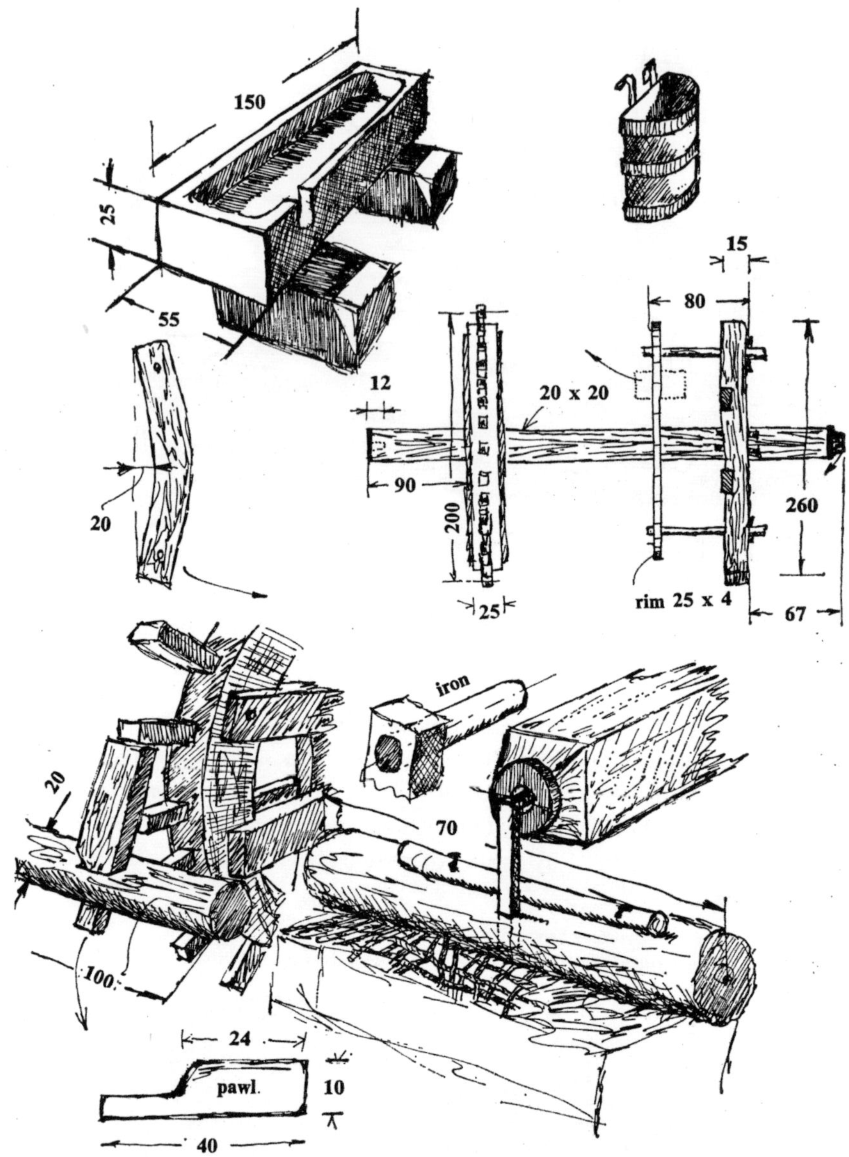

Fig. 24. Details of parts of the sāqiya at Mi'timdiyya. Stone trough. Bucket for bucket-chain wheel. Pawl mechanism. Bearing for the horizontal shaft. Drawn in Egypt 1964.

Passing now to the horizontal shaft, we reach the little cogwheel, which is a copy of the large wheel, so a detailed description is unnecessary. This brings us to the potgarland wheel itself, which consists of more than 25 components. To understand this complicated wheel, the reader is advised to study Fig. 23, on p. 29.

The rim facing us has no spokes so it appears to be suspended in space. In fact, the trough inside the wheel is supported by two small cantilever beams. There is thus no room for spokes in the wheel as they would traverse the trough supports. In this wheel, too, the earthenware pots have been replaced by the buckets shown in Fig. 24. The buckets are tied to two small pegs lashed to the potgarland rope.

The trough is made of concrete and its outlet is placed a little way up so there is always a certain depth of water in the trough, thereby minimizing splashing. From the trough the water flows out to the fields through an underground pipe.

The pawl

On this journey through the Egyptian sāqiya we must not forget the pawl. See Fig. 24. This important device was described in the section on the Spanish noria, and it will be remembered that a defective pawl may result in the draught animal being killed. In the Egyptian sāqiya, there is the

Fig. 25. Sāqiya with bucket-chain wheel. Common in the Fayyūm, Egypt.

added danger that the guide-boy who sits on the big wheel may get knocked off it if the wheel suddenly reverses. There is a Coptic story about a little boy who was pushed off the wheel and into the well by the Devil. W. Spiegelberg has translated the story as follows (cf. p. 111):

> .... ein Knabe war auf ein Wasserrad [no, this must be the horizontal cog-wheel] gestiegen, indem er das Vieh des Klosters antrieb. ... Da stieß der Teufel den Knaben in den Brunnen, und er ertrank in dem Wasser.

I assume that the Devil must have removed the pawl, which would be quite simple as this is loosely placed in a square hole.

Lastly, let us take a look at the iron hook on the long palm beam, which has no technical purpose, but which is, all the same, a very important piece of equipment. The guide-boy's bread-bag is hung from this hook, well out of the reach of hungry dogs. See Figs. 21 and 25.

Type *IB2b*
Fayyūm

This is the most effective type of wheel for small lifts, and it is very common in the Fayyūm district of Egypt. See Fig. 25, on p. 31.

Fig. 26a. Noria at Formia between Rome and Naples in 1971.

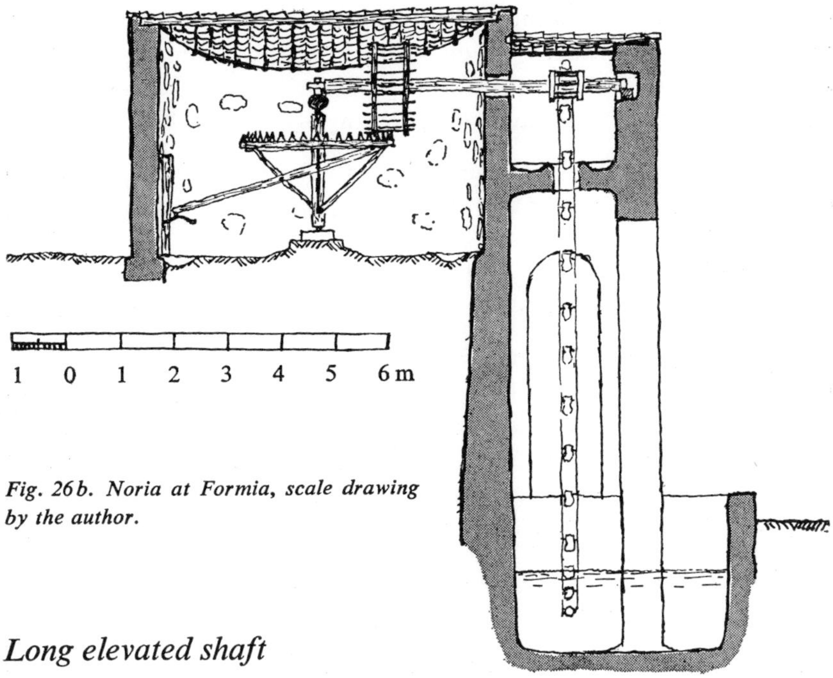

Fig. 26b. Noria at Formia, scale drawing by the author.

Long elevated shaft

The next four types belong to the category with an elevated shaft. They are now a rarity in the field of irrigation, but were not unknown a hundred years ago.

Type *IC1a*
Formia, Italy

During a visit to Italy in 1971, I found, quite unexpectedly, a sāqiya of the elevated shaft type in the town of Formia, between Rome and Naples. Although the machinery had obviously not been used for some years, it was in good condition, except that the potgarland was missing. See Figs. 26a and 26b.

Type *IC1b*
Palestine

This type was once common in Palestine but has now (1969) completely disappeared. As will be seen from Fig. 27, the lantern pinion is carried by

Fig. 27. Sāqiya with elevated shaft. This construction was once common in the Palestine. The long diagonal bar forms part of the pawl mechanism.

the horizontal shaft, and this arrangement is also shown in a rather poor photograph in a German work on Palestine, Gustaf Dalman: *Arbeit und Sitte in Palästina*, Vol. *II*, *Der Ackerbau*. (Schriften des Deutschen Palästina-Instituts, Vol. 5, (Gütersloh 1932)), fig. 47 and pp. 225–226).

Type *IC2a*
India

We have many pictures of sāqiyas from India, most of which are iron constructions, with a long horizontal shaft, but Fig. 28 shows the main outlines of a wooden sāqiya. The peculiar structure on the right of the men is a cog-wheel with 12 cogs. The potgarland wheel has a number of details that are reminiscent of the wheel from West Pakistan. See Fig. 18a.

Type *IC2b*
Syria

This Syrian type of wheel, which has an elevated shaft and radial cogs, in combination with a bucket wheel, I know only from sketches in a French work, Jacques Weulersse: *Le pays des Alaouites*. (Institut français de Damas). Tours 1940, fig. 82, and from a photograph, A. J. Jaussen et R. Savignac: *Mission archéologique en Arabie. I.* Texte et atlas. (Publications de la Société des fouilles archéologiques). Paris 1909, fig. 31, p. 32.

The last part of this section, group *I*, is devoted to a number of types that do not fit the classification.

Type with hollow rim
Ismailia

For lifts of at the most one metre, it is sometimes found advantageous to use a large wheel with a compartmented rim instead of a potgarland. A sketch of such a wheel, which I found in ruins at Salḥin near Aleppo,

Fig. 28. Sketch of sāqiya with elevated shaft and cog-wheel. The cog-wheel has only 12 cogs. The circular ox-track is below ground level. A long iron-plate chute leads the water to the field. This gutter is placed over the circular track. India about 1950.

Fig. 29. Defective noria at Aleppo. The waterwheel has hollow rim, with about 16 compartments. The upright shaft is held in position by two beams, weighted by a couple of large stones, to hold them against the piles of stone on which they rest. The centrifugal pump in the foreground was intended to replace the noria, but this, too, is out of order.

is shown in Fig. 29, and there is a photograph of one in a book on water problems in Syria, Nazim Moussly: *Le problème de l'eau en Syrie.* Lyon 1951, plate *III* C.

The best known of the machines based on this principle are to be found in the Nile Delta. The design, which is shown in Figs. 30 and 31, differs in many ways from that of the ordinary sāqiya. The well is located within the circular ox-track, and instead of the long beam, a short pole is used.

Fig. 30. Sāqiya with compartmented rim. The correct term is tābūt. Survey carried out in Ismailia in Egypt 1964.

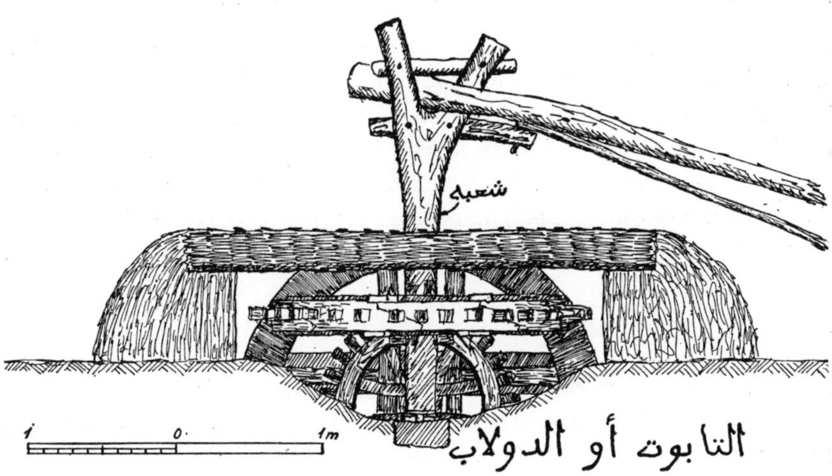

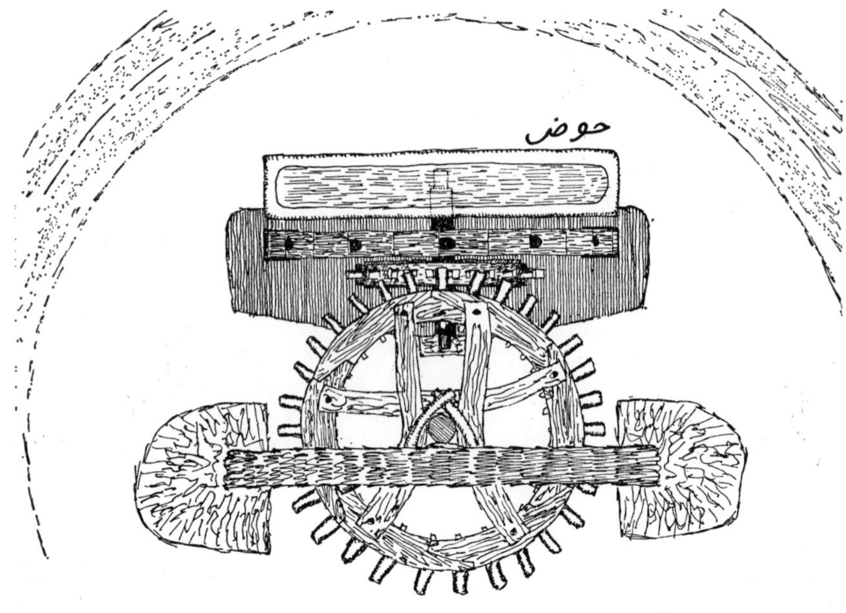

Fig. 31. The same tābūt as Fig. 30, but seen from above. This type of machine is very popular in the Nile Delta and must be the successor of the Roman invention depicted in the wall painting in Alexandria. See Fig. 108 on p. 152.

The compartmented wheel is arranged in direct connexion with the cog-wheel. The rim itself consists of 16 compartments, each with two apertures. There are literally thousands of this type of machine in the region north of Cairo. Some Egyptians call it a sāqiya, but its real name is tābūt. I have found illustrations of a similar construction from Italy in K. Jaberg und J. Jud: *Sprach- und Sachatlas Italiens und der Südschweiz.* Illustrationsband, von Paul Scheuermeier: *Bauernwerk in Italien und der rätoromanischen Schweiz.* Erlenbach-Zürich 1945, p. 240.

Type with paddlewheel
Nile Delta

Another machine common in the Nile Delta operates with a paddle-wheel, in which the longitudinal section of the blades resembles a spiral. The paddlewheel, which is always made of iron nowadays, is referred to and illustrated in an F.A.O. publication F.A.O. Agricultural Development Paper No. 90, *Water-Lifting Devices for Irrigation*, prepared by A. Molenaar, Rome 1956. The design principle is a very old one, as is illustrated by a miniature from the 12th century. See Figs. 54a and 54b.

Group *II*. Machines without gear

As indicated in the foregoing, these wheels can be classified in four groups: (A) water-driven wheels, (B) treadwheels, sub-divided into (1) pot-wheels and (2) compartmented wheels.

Fig. 32. Hydraulic noria from Hīt on the Euphrates. The wheel is provided with a number of mats that act as paddles in a waterwheel. An endless row of slender pots is fastened to the rim. See also Fig. 71. The drawing shows only one wheel, although there were actually five more behind it.

A. Water-Driven Wheels

Type *IIA1*
Hīt, Iraq

Fig. 32 shows a type of wheel that is still (1969) in use at Hīt on the Euphrates in Iraq. The construction actually has five wheels working on one shaft, but for purposes of clarity, the rear four wheels have been omitted from my drawing.

A large number of pots are lashed to the outer rim. The pots begin to empty as they reach the highest point of the wheel but are not completely drained until they are well on the way down again.

Type *IIA2*
Ḥamā, Syria

The famous wheels of Ḥamā are of the compartmented type. The rims of the largest wheels consist of 120 compartments. These wheels, nine of which were still in operation in 1969, although mostly as a tourist attraction, provide a fascinating spectacle, and the experience is enhanced by the "organ music" produced by their wooden bearings. This music is a mixture of noise and true notes, the deepest of which lie in the frequency band 120–170 Hz, or nearly two octaves below concert pitch.

The largest of the wheels was surveyed in 1935 by Ejnar Fugmann, a Danish architect. Today (1969), this wheel lies partly in ruins. The drawing in Fig. 1 was made by Fugmann on the basis of his extremely complicated survey, which necessitated more than 275 different measurements. Shortly before he died in 1965, Ejnar Fugmann requested me to publish this survey.

The extreme complexity of the design is apparent from Fig. 33 and the construction is as follows: first, four continuous beams constituting the four double spokes are mounted on either side of a frame wedged tightly in position round the hub. Next, the inner rim is constructed and, at the same time, the fan-shaped spokes are nailed in place. The wheel has to be turned at intervals during assembly in order to avoid possible wobbling.

Fig. 34. Ḥamā wheel. The rim consists of 120 compartments, end to end. The paddles are fixed between the compartments.

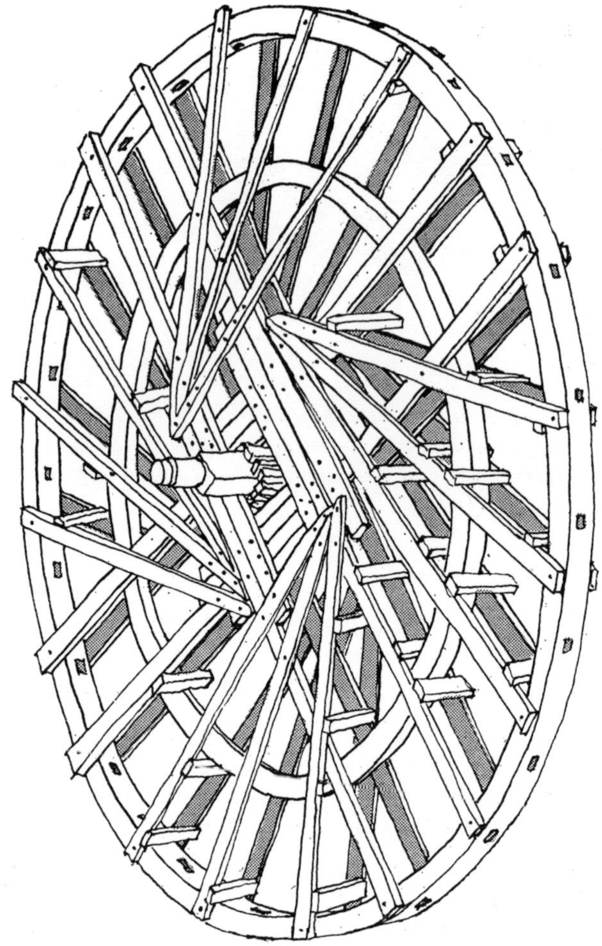

Fig. 33. Model of a Ḥamā wheel. The outer compartmented rim has been omitted. The drawing is not a true representation, the scale of the timbers being out of proportion to the overall size of the wheel. See Fig. 1 on p. 8.

Finally, the outer rim for the compartments is mounted. The design of the compartments is shown in Fig. 34.

Before leaving these wheels, there are a few points with which I think the reader should be acquainted.

A great many authors have confused these wheels with the geared noria, presumably because both types lift water and both are called a noria. Therefore, in order to differentiate clearly between them, the Ḥamā-type of noria will be termed a water-driven or hydraulic noria, and the other type, an angle-gear or animal-powered noria. Julio Caro Baroja has avoided the pitfall of ambiguity by devoting a separate paper to each type, classified above as groups *I* and *II* (cf. p. 11).

Unfortunately, there are a number of intermediate forms, classified here as "other geared machines", and this has necessitated treatment of all the various types of wheels. The great wheels at Ḥamā have therefore been included in this book even though, in principle, they belong to a group that falls outside its scope. Had the intermediate types been of recent date, it would have been possible to omit the hydraulic noria altogether, but the very oldest picture we have of a sāqiya shows a type combining a gear, oxen, and a waterwheel of the Ḥamā-type.

Classification and Technology

B. Treadwheels

Type *IIB1*
Ríotinto

I have not succeeded in finding a specimen of a treadwheel with pots, and although such undoubtedly exist, they must be exceedingly rare. In the Ríotinto mines in Spain there were probably such wheels. See G. C. Boon and C. Williams, p. 124, note 25 (cf. p. 160).

Type *IIB2*
Ostia Antica

There are both recent and ancient examples of the treadwheel-compartmented type, the design of which is shown in Fig. 35. In connexion with the treadwheel, we encounter the serious problem of stability, or the ability of the treadwheel operators to control the movement of the wheel. There are two types of treadwheels, one in which men turn the wheel by treading on steps inside the wheel, and one in which the men tread the wheel from above. Since antiquity, treadwheels in which the men work

Fig. 35. Roman treadwheel with compartmented rim from Santo Domingo in Portugal. The remains of wheels of this type are to be found in the British Museum, Musée de Conservatoire national arts et métiers in Paris, and in el Museo de Ríotinto in Spain.

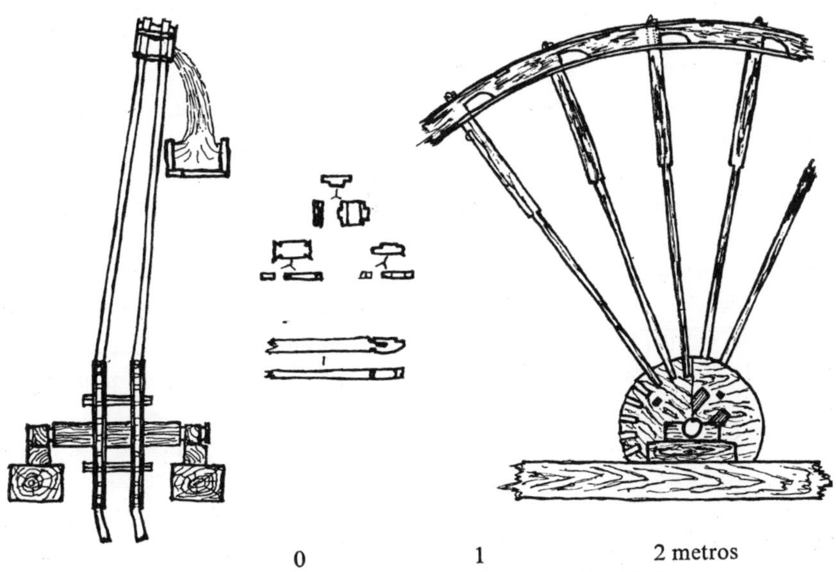

inside the wheel have been used to operate winches. In a comment to this, F. R. Forbes Taylor writes:

> With the men inside, it was almost impossible for them to lose control, even without the use of brake or pawl ....

F. R. Forbes Taylor: *The Winch from Well-head to Goliath Crane.* [The Chartered Mechanical Engineer (April 1962), p. 203].
The treadwheel operated by men treading at the top is regarded as an unstable system, although bearing friction improves the stability. Moreover if the wheel is provided with a pawl, I should think that the men have sufficient control over it for water-lifting purposes.

At the end of this section on machines in group *II* we now come to "other types of machines without gear."

Types of wheel turned by hands and feet

The treadwheel can also be coupled to a paternoster wheel. An example of such a combination is shown in Fig. 95. Finally, there is another type of treadwheel, which is turned by means of both hands and feet, and which is also designed to carry a potgarland. See Figs. 56 and 57.

Terminology

The problem of the terminology of waterwheels is practically insoluble. In a geographical region in which a number of different types of machine are used, as in Northern Egypt, the people use a generic term to cover them all. Dictionaries are sometimes of help, but the translations are not always reliable.

In the following a number of the most common terms in use today will be discussed.

In Spain, the term noria is used both for water-driven wheels and animal-powered wheels. To avoid misunderstandings, Caro Baroja uses two terms, *la rueda de corriente* and *la noria de sangre*. In Ibiza, the term, *sèni*, which is of Arabic origin, is used.

In Syria, the people seem to be more consistent. Only the great wheels at Ḥamā are termed norias or, more correctly, the classical form, nāʿūra. This term, on the other hand, is unknown in Egypt, although Europeans use the Spanish form, noria, when referring to the Egyptian sāqiya. In the vicinity of Aleppo, the old Persian word, dawlāb, which is well known in Egypt, the Sudan and Iraq, is used. The term does not refer to a particular type, but is simply a general expression for wheel, possibly a waterwheel.

In Syria, the term ġarrāf is used for the geared noria, whereas the word madār simply means mill or circular track.

In addition to the terms commented on above, there are other terms, such as ṭanbūša, ṭabliyya and ṭulunba. They can hardly be translated as anything but water-lifting wheels.

Bibliography

Each of the works referred to below devotes a few pages to the terminology of the various machines:

Caro Baroja, Julio. *Norias, azudas, aceñas.* [Revista de dialectología y tradiciones populares, Vol. *X* (Madrid 1954), pp. 29–160].

– *Sobre la historia de la noria de tiro.* [Revista de dialectología y tradiciones populares, Vol. *XI* (Madrid 1955), pp. 15–79].

Colin, G. S. *La noria marocaine et les machines hydrauliques dans le monde arabe.* [Hespéris, Vol. *XIV*. (Paris 1932), pp. 22–60].

Dalman, Gustaf. *Arbeit und Sitte in Palästina.* Vol. *II Der Ackerbau* (Schriften des Deutschen Palästina-Instituts, Vol. 5 (Gütersloh 1932), pp. 225–228).

Dias, Jorge y Fernando Galhano. *Aparelhos de elevar a água de rega,* Porto 1953.

Littmann, Enno. *Die Sāqiya.* [Zeitschrift für ägyptische Sprache und Altertumskunde. Vol. 76 (Leipzig 1940), pp. 45–54].

Mitwally, M. *Native Methods of working a Sakia in the Northern Province of the Egyptian Sudan.* [Bulletin of the Faculty of Arts, Vol. 13, part *II* (Cairo 1951), pp. 61–66].

Nicholls, W. *The Sakia in Dongola Province.* [Sudan Notes and Records, Vol. 1 (Khartoum 1918), p. 23].

Nicholson, H. A. *Saqia Terminology in Dongola.* [Sudan Notes and Records, Vol. 18, part 2 (Khartoum 1935), p. 314 ff].

Schäfer, Heinrich. *Nubische Texte im Dialekte der Kunūzi.* [Abhandlungen der Königliche Preußischen Akademie der Wissenschaften, Philologisch-historische Klasse nr. 5 (Berlin 1917), p. 107 ff].

Schiøler, Th. *Las norias ibicencas.* [Revista de dialectología y tradiciones populares, Vol. *XVIII* (Madrid 1962), pp. 480–486].

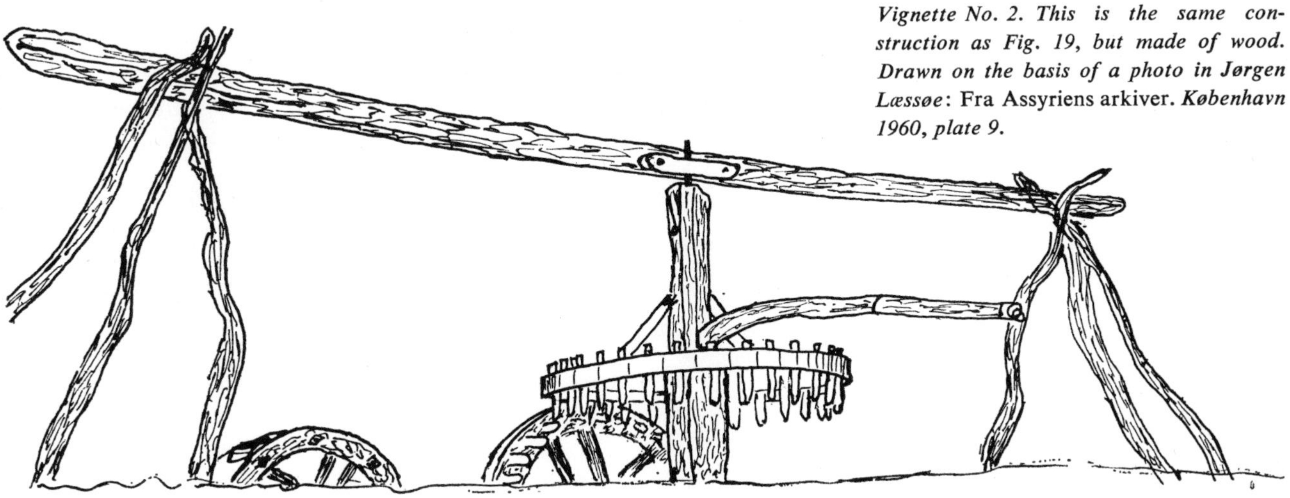

Vignette No. 2. This is the same construction as Fig. 19, but made of wood. Drawn on the basis of a photo in Jørgen Læssøe: Fra Assyriens arkiver. *København 1960, plate 9.*

CHAPTER 2

The European Paternoster Pump

Literature from A.D. 1400 to A.D. 1900 reviewed in inverse chronological order

Introduction

At least three publications of recent date are of particular value as a starting point for a study of the history of water-raising contrivances.

The first of these is Joseph Needham's *Science and Civilisation in China,* Vol. 4, part *II.* Mechanical Engineering. Cambridge U. P. 1965, pp. 330–362. The Chinese literature on water-raising contrivances is very extensive and allows us to trace their history far back in time, but the point of significance in this study is that machines with a right-angle gear did not appear in China until the 9th or 10th century. Needham's book contains a large number of illustrations and many interesting references to mediaeval literature and, at the end, some new references to Indian literature.

The second study is by Julio Caro Baroja: *Sobre la historia de la noria de tiro.* (The History of the Animal-Propelled Noria) [Revista de dialectología y tradiciones populares, Vol. *XI* (Madrid 1955), pp. 15–79]. This richly illustrated work is the only monograph on the geared noria. It is divided into seven chapters: 1) Introduction; 2) The geared noria in works of mediaeval Arab authors; 3) The geared noria, its expansion in the Islamic world; 4) The geared noria in India; 5) The geared noria in China; 6) Returning to the west; and 7) The geared noria in Spain.

The first few pages probe the origin of the geared noria. The question is whether the inventor is to be sought in ancient Egypt or Mesopotamia, or further east, perhaps in India or China. Caro Baroja finds these possibilities unlikely and turns to the Greek papyri from Egypt:

> *Creo que no hay* ... I do not believe there is sufficient evidence to prove that the word *mēchanē*, used in the papyri is synonymous with sāqiya.

He continues in this vein:

> ... as regards the classical texts, these contain no descriptions of the geared noria.

We shall return to these two views later.

The chapter on Spain is particularly valuable because it contains a large number of references to old Spanish literature.

The third book is R. J. Forbes: *Studies in Ancient Technology*, Vol. *II*. 2nd ed. Leiden 1965. This major work on ancient technology is an invaluable aid to further study because of its comprehensive bibliography. The chapter, 'Methods of Raising Water' begins on page 32, but the reader would be well advised to study first the classification given on page 39. Forbes classifies all known water-raising contrivances in three groups: A) Discontinuous water supply; B) Semi-mechanical; C) Continuous water supply by water-lifting machinery. The last group, C), is sub-divided according to the type of motive power: a) Moved by animals or men; b) Moved by the river; and a) is further divided into: 1) Archimedean screw; 2) Wheel of pots; 3) Compartment wheel (hollow wheel); and 4) Chain of pots (potgarland).

As will be seen from the above, Forbes does not differentiate between geared and ungeared wheels, and this, combined with the fact that he uses both tābūt and sāqiya as terms for non-geared wheels, may tend to confuse the reader.

The period A.D. *1400 to* A.D. *1900*

There is a great deal of literature shedding light on the geared noria during this period, but only books containing technical information are of interest in the present context. The following references are arranged in inverse chronological order to allow the reader to form a rough idea of the development in the field of technology. I have allowed this consideration to take precedence over the fact that the method forces us to move unsystematically from one country to another.

1894 *Russia*

There are very few material sources from Southern Russia, but a single item of interest is to be found in Laufer (1933), p. 242 (cf. p. 84), who cites Henri Moser: *L'Irrigation en Asie centrale*, Paris 1894, p. 266:

> He observed these tshigir (sāqiyas) on the banks of the Amu Darya and in Khiwa (Turkmenistan).

From the subsequent description it appears that the author was referring to a geared noria, turned by a horse, a camel or an ass, which could raise 4–5 cubic metres per hour.

1889 *Egypt*

The longest and probably the best description of the Egyptian sāqiya was written by the Frenchman Barrois, whose book was later translated into English. I. Barrois: *Irrigation in Egypt*, Washington 1889, pp. 77–78. After discussing the various devices, Barrois studies their efficiency. On the basis of numerical data from *Description de l'Égypte* (cf. p. 45), he computes their output and compares this with the theoretical work capacity of European draught animals. Barrois does not take the

Fig. 36. Sāqiya in the province of Sind, West Pakistan.

last step – that of calculating the ratio between these two values in order to determine the efficiency, but one gets the impression that he is perfectly conversant with this line of reasoning; at any rate, he expresses dissatisfaction over the fact that only a third of the input is actually utilized.

1868 *India*

B. H. Powell: *Hand-Book of the Economic Products of the Punjab*, Vol. *II*, Roorkee Lahore 1868–1872, pp. 207–208.
There is a considerable number of sāqiyas in Northern India, but they are all made of iron. According to Mushtaqur Rahman: *Kunstvanding med øseværket Nar i Vest-Pakistan.* [Kulturgeografi nr. 99 (Aarhus 1966), pp. 325–330], there are at least 100,000 sāqiyas in West Pakistan alone. The older literature contains descriptions of wooden sāqiyas; Laufer 1933, p. 244, cites from Powell one text that seems to relate to a contrivance more or less similar to the machine shown in Fig. 36.

1842 *USA*

Thomas Ewbank: *A Descriptive and Historical Account of Hydraulic and other Machines for Raising Water.* N.Y. 1842.
This American work is often cited, but much of it relies more on a vivid

imagination than on scientific investigation. For example, in a passage on *The Hanging Gardens of Babylon*, the author writes:

> There is a machine noticed by ancient authors (cf. p. 163), which probably belongs to this part of our subject ... p. 133. As no account of the nature of this machine has been preserved, we are left to conjecture the principle ... p. 134.

1822
Egypt

Napoleon Bonaparte [C. L. F. Panckoucke]: *Description de l'Égypte* ... Paris 1809–1822, 2nd ed. 1821–1829.
This comprehensive French work, which is credited to Napoleon Bonaparte, is a valuable source of information. It includes several elephant folios of drawings, two of which are entitled *État moderne*. The second half of volume *II* has the sub-title, *Arts et métiers*, and volume 12 of the main work gives an *Explication des planches des arts et métiers*. (Paris 1823).

One of the few investigations of the Egyptian sāqiya was carried out by an engineer named Faye, and his data are to be found in Vol. 18, Part 2, 2nd ed. of the above work (published in Paris 1829), facing p. 544.

Faye ran trials on seven different machines, some of which were used for irrigation and others for domestic purposes in Alexandria. As draught animals he mentions the horse, the ox and the ass, which makes three to five rounds a minute. Faye measured, among other things, the number of pots and the time it took for the potgarland to complete one revolution. Unfortunately, Faye does not say how many oxen, whether one or two, were harnessed to each machine; in other words, we do not know the tractive force or the velocity. It appears from the detailed calculation and from the description that Faye counted the number of pots passing per minute and multiplied this by the volume of a single pot, thereby arriving at the discharge of the machine. He used a standard 1.6 litre pot to calculate the flow of the seven different machines, which implies that he assumed all of them to be equipped with the same size of pot. I think Faye overlooked the fact that the volume of the pots varies greatly from one machine to another, and it is possible that some of his results should really be multiplied by a factor of four or five.
In the text, Faye estimates the mean discharge to be 72 litres per minute but omits to mention the head. In some of the tests, the lift was 5 metres and in others 10 metres. It is not my intention here to prove that Faye made mistakes, but simply to give an example of how difficult it was, only 150 years ago, to analyse a machine theoretically.

One of the machines included in the investigation is located at a public baths, and another in the big mosque. The text also mentions a large water tank supplied by 70 (?) sāqiyas, each operated by two men and three oxen, working round the clock, the animals being changed six times a day.

We shall now turn to the illustration in Napoleon Bonaparte's *Description de l'Égypte, État moderne*, 2 volumes in elephant folio, Paris 1822:

Vol. *I*, plate 73, figs. 1, 2, 3 and 4. One of the hydraulic works frequently mentioned in the literature is located in the Citadel in Cairo. It has been drawn to scale several times, and we shall return to it later.
Vol. *I*, plate 78, fig. 1. A potgarland wheel with potgarland can be glimpsed in this illustration. It is apparently of the hand- and foot-operated type, as Fig. 56.
Vol. *II*, plate *III*, figs. 1, 2, 3, 4, 5 and 6. This plate shows a really good drawing of a tābūt, and a number of details indicate the hollow-rim system. See also Figs. 30 and 31.
Vol. *II*, plate *IV*, figs. 1, 2, 3, 4 and 5. This shows a scale drawing of a sāqiya. The drawing is a fine work of art, although unfortunately not flawless; the pots are shown upside-down and the pawl mechanism has escaped the artist's attention.
Vol. *II*, plate *V*. This drawing has been copied several times. It shows two sāqiyas in operation at the same well. On the left side of the illustration there is an Egyptian smoking a hookah. The engraving has been reproduced in innumerable books, but unfortunately cut down in such a way as to give the impression that there is only one sāqiya connected to two potgarland wheels. Pemp (cf. p. 149) did not discover this error, and on page 9 of his dissertation, the drawing is used to prove the existence of a water-lifting machine turning two paternoster chains. Caro Baroja (1955), p. 34, also sees two potgarlands. The mistake can be traced back to Moritz Rühlmann: *Allgemeine Maschinenlehre*. Braunschweig 1875, Vol. 4, p. 559, fig. 383, although he did indicate that he had reduced the drawing.

1786
Spain

Joseph Townsend: *A Journey Through Spain in the Years 1786 and 1787; with Particular Attention to the Agriculture, Manufactures, Commerce, Population, Taxes, and Revenue of that Country*; *and Remarks in Passing Through a Part of France*. London 1791, in three volumes.

After a short stay in Paris, we meet Mr. Townsend in Narbonne. (Vol. *I*, p. 79):

> At Narbonne there is a little stream, which, by the industry of the inhabitants, proves to them a more certain source of wealth than if its sands were gold.
>
> This stream empties itself into a canal of more than half a mile in length, by the side of which they have extensive gardens, watered from it by hydraulical machines of remarkable simplicity.
>
> They consist of a vertical wheel of twenty feet in diameter, on the circumference of which are fixed a number of little boxes, or square buckets, to raise the water out of a well, communicating with the canal below, and empty it into a reservoir above, placed by the side

> of the wheel. These buckets have a lateral orifice to receive and to discharge the water. The axis of this wheel is embraced by four small beams, crossing each other at right angles, tapering at the extremities, and forming eight little arms. This wheel is near the centre of the horse walk, contiguous to the vertical axis, into the top of which the horse beam is fixed; but near the bottom it is embraced by four little beams, forming eight arms similar to those above described, on the axis of the water wheel. As the mule, which they use, goes round, these horizontal arms, supplying the place of cogs, take hold, each in succession, of those arms which are fixed on the axis of the water wheel, and keep it in rotation.

It is easy, from this masterly description, to form a picture of the type of noria in question, which is of the tābūt-type. The cog-wheel gear is of the type shown in Fig. 15, but the waterwheel mechanism is of the compartmented rim type. I have seen the ruins of such a construction outside Aleppo in Northern Syria. Fig. 29 gives an impression of the design of the wheel.

We may wonder in what country this design originated, but before attempting to answer this question, I should like to return for a moment to Mr. Townsend's report:

> This machine, than which nothing can be cheaper, throws up a great quantity of water, yet undoubtedly it has two defects: the first is, that part of the water runs out of the buckets and falls back into the well after it has been raised nearly to the level of the reservoir; the second is, that a considerable proportion of the water to be discharged is raised higher than the reservoir, and falls into it only at the moment when the bucket is at the highest point of the circle, and ready to descend.
>
> Both these defects might be remedied with ease, by leaving these square buckets open at one end, making them swing on a pivot, fixed a little above their centre of gravity, and placing the trough of the reservoir in such a position as to stop their progress whilst perpendicular, make them turn upon their pivot, and so discharge their contents.

The two problems raised by Mr. Townsend are of real importance. Some of the water never reaches the trough and some of it is raised unnecessarily high, both factors reducing the efficiency. Mr. Townsend may possibly have seen the solution he mentions in B. F. de Bélidor: *Architecture Hydraulique*; *ou l'Art de Conduire, d'Élever et de Ménager les Eaux* ... 4 vols. Paris 1737–1753. I myself have used a later edition of this work. The pivoted boxes mentioned in the above extract are shown in a drawing in Vol. 1, 1782, Part 1, Chapter 4, plate 9.

Twenty kilometres from Barcelona, Mr. Townsend again encounters the noria, but this time of another type, p. 103:

> ..... Every farm has its Noria, a species of chain pump, which, from its extreme simplicity, seems to have been the invention of the most remote antiquity. By means of this machine they every morning draw a sufficient quantity of water from the well for the service of the day, and in the evening distribute it to every quarter, according to the nature of their crops.

Not until page 170 do we find a description of this noria, which proves to be of the type with a short axle and a lantern wheel of type *IA1a. Ibiza.*

The following lines are a slight digression from the subject of the present thesis, but I have included them as an illustration of the thinking at that time.

Mr. Townsend recalls a story he has heard in connexion with a chain pump, which is, in principle, reminiscent of the Spanish noria. A chain pump is a bilge pump and was used in His Majesty's Navy. The system is based on the use of an endless chain to which a large number of leather pistons are attached at suitable intervals. As the chain and pistons are drawn up through a tube, each piston lifts a quantity of water.
However, a new design had been invented, which the Admiralty had seen demonstrated and had approved. A number of ships had been fitted with the new chain pump, but on the way back from the West Indies, the H.M.S. Centaur went down because her new pump had become clogged with sand and thus stopped. But worse still, it seems that the inventor had attached a report to his invention, showing that the new pump could raise much more water with the aid of only two men than could the old chain pump with seven. Mr. Townsend doubts "that this trial was conducted fairly". We can read between the lines that he is not only indignant, but furious.

In Barcelona, Mr. Townsend meets some gentlemen who have tried to improve the noria. The Spanish noria has two cog-wheels, but the new design has four. Mr. Townsend's comment on this is to be found on p. 176:

> The movements in this machine are too complicated, and thereby both the expense and the friction are increased.

He then turns to the drawbar, which he considers should be arranged in another way:

> .... the horse walk [diameter] is too small, and the beam [drawbar] being behind the horse, instead of being placed over his shoulders, the line of draught makes with it an angle of forty-five degrees, and thereby one half of his force is lost.

The solution proposed by Mr. Townsend is known on Majorca. Here, the

Fig. 37 and Plate 5. Copper engraving of a sāqiya in Frederic Louïs Norden: Voyage d'Égypte et de Nubie. *It will be noticed that the little cog-wheel, which is half-hidden behind the big cog-wheel, is shown too high. If it were really located here, the ox would have to jump over the horizontal shaft. Compare this drawing with the original on the the next page.*

two horizontal traces are replaced by two vertical bars, and the drawbar is raised so that it passes over the shoulders of the animal.
The whole description gives an excellent impression of the activity that took place in the field of technology in Europe at that time. The similarity between the noria seen by Mr. Townsend in Narbonne and the one I have seen in Aleppo is not fortuitous; it is simply a case of a design imported from the Orient many centuries ago.

"In future, the French Academy will not accept any designs for *perpetuum* 1775
mobile." This piece of information becomes of interest when seen in the *Paris*
light of the following pages.

One of the best scale drawings of a sāqiya and a tābūt was made two 1774
hundred years ago and published some years later by Carsten Niebuhr: *Egypt*
Reisebeschreibung nach Arabien und andern umliegenden Ländern, Vol. *I*, Kopenhagen 1774, p. 148 and plate *XV*. This drawing is also included in Thorkild Hansen's enthralling book, *Arabia Felix*, N.Y. 1964.

Several Danes have found the sāqiya a fascinating subject for study; 1755
as early as 1755 we find a couple of drawings in Frederic Louïs Norden: *Egypt*
Voyage d'Égypte et de Nubie, Vol. *I*, Copenhague 1755.
Plate *XIX* contains a scale drawing of Joseph's Well in Cairo. I have

Fig. 38 and Plate 6. Frederic Louïs Norden's own drawing. The shaft of the little cog-wheel disappears under the circular ox-track, which is correct.

Fig. 39. Eighteenth-century engraving of a rāqiya, which, however, is totally unsealistic.

myself had the opportunity of inspecting this well, and, apart from the fact that the machinery is broken and out of action, the place has not changed since Norden visited it. The history of the well can be traced back to the 11th century.

The drawings of *Différentes Machines Hydrauliques* in plates *X*, *XL* and *LIII* are unfortunately not quite perfect, but reproduction techniques at that time did not permit the use of Norden's own drawings. However, some of these are under the trusteeship of the Royal Library in Copenhagen, Codex G.k.S. 447 *I* and *II*, and others are in the possession of *Det kongelige danske videnskabernes Selskab*.

In plate *XL*, see Fig. 37, there is one glaring error, which does not occur in the original drawing. See Fig. 38. If, in our mind's eye, we watch the ox walking round, we see that it will have to jump over the horizontal shaft in order to get past it. The original drawing relieves the poor animal of this necessity.

1738 *Egypt*

There are other examples of artists allowing their imagination to run away with them, for instance, in Th. Shaw: *Travels, or Observations Relating to Several Parts of Barbary and the Levant*. Oxford 1738, facing p. 336, where a sāqiya is shown. See Fig. 39. Apart from the lantern wheel, there is hardly anything correct in this drawing. On the other hand, there are some interesting remarks on page 431 concerning water tanks in connexion with sāqiyas:

> When their Melons and Sugar Canes require to be refreshed, they strike out a Plug, that is fixed in the Bottom of one of these Cisterns; and then the Water gushing out ....

This is one of the few references to cisterns, or water tanks, in connexion with irrigation contrivances. In the Greek papyri, to which we shall revert, the term *lakkos* is frequently encountered in relation to water-lifting machines. It means a hollow, but the question is whether it also means an irrigating cistern or a watering trough for the oxen.

The European Paternoster Pump

Renaissance

There is a vast European literature on machines from the years 1500 to 1750. I have examined some works, but have had to pass over others. Readers interested in a study of this period are referred to Eugene S. Ferguson: *Bibliography of the History of Technology*. (The M.I.T. Press) Cambridge, Mass., 1968.

1737
France

In the work by Bélidor mentioned earlier (cf. p. 47), there is a description and a large drawing of an enormous paternoster machine dating from 1722. Bélidor ...., Vol. 1, Paris 1782, Part 1, pp. 370–374 and plate 5. This machine has three bucket chains and a total of seven cog-wheels. Bélidor is dissatisfied with the arrangement because he feels that a machine with four horses should have a higher output. The entire pump has a discharge of 44 cubic metres per hour with a lift of 8 metres. A comparison of this output with that of present-day norias presents a favourable picture, and Bélidor had no real grounds for dissatisfaction. On the basis of his knowledge of the power capacity of a horse (la quantité de mouvement), Bélidor arrives at an efficiency of 0.3 for the pump. He makes the error of assuming that a horse on a circular track exerts the same power as a horse harnessed to a carriage, and thus arrives at this rather poor figure. The important thing in this context is, however, that Bélidor was the first writer to mention the efficiency (l'effet de cette machine) of a paternoster pump.

1724
Germany

Jacob Leupold: *Theatrum Machinarum Hydraulicarum*, Vol. *I*, Part 1 Leipzig 1724. This work contains an engraving of a sāqiya with an elevated shaft.

1690
Poland

Stanisław Solski: *Architekt Polski*, Krakau 1690, reprinted, Wrocław (Warsaw) 1959. Here, too, there are a couple of sketches of paternoster pumps, but they are copies from far older works.

1661
Germany

Georg Andreas Böckler: *Theatrum Machinarum Novum, Schauplatz der mechanischen Künsten Mühl- und Wasserwerk*, Nürnberg 1673 (1661). In this now very rare book there are more than 20 large copperplates of paternoster pumps, each more fantastic than the other. All except one, fig. 133, are pure figments of imagination, some of which Böckler borrowed from other authors, for example, Agricola and Zonca. There are two points of particular interest in these drawings. First, we find, throughout, the use of a bucket-chain wheel of the same design as is found in Syria, and secondly, some of the imaginary machines are driven by weights. We encounter just such a hydraulic machine from the realms of fantasy in an Arabic manuscript from the 11th or 12th century. See Schmeller, pp. 10–12 (cf. p. 73). One of Böckler's more credible drawings has been reprinted in Martin Jensen: *Civil Engineering around 1700.* (Privately printed). Copenhagen 1969, p. 114.

1643 *Poland*

The Travels of Peter Mundy IV. Works issued by the Hakluyt Society, 2nd Ser. No. *LV.* London 1925, p. 203. In Warsaw, Peter Mundy is given a description of a paternoster wheel that draws water up – by itself. Half of the water runs to a palace and the other half drives the gravity wheel. Unfortunately for posterity, Peter Mundy's technical description is incomprehensible, but he continues:

> .... soe thatt if this bee true, as I thinck itt is, itt may bee rightly called a perpetual Motion.

However, by the time Peter Mundy sees the machine, the original design has been changed and it is now driven by twelve slaves.

1607–1614 *Germany*

Heinrich Zeising: *Theatrum Machinarum, so in sechs Theil bestehend...* Leipzig 1607–1614. – A later edition, from 1708, fig. 4, p. 18, shows a high-lift sāqiya.

1607 *Italy*

Vittorio Zonca: *Nuovo Teatro di Machine et Edificii.* Padua 1607. Reference is frequently made to this book, for example, in F. M. Feldhaus: *Die Technik der Vorzeit, der geschichtlichen Zeit und der Naturvölker.* 2nd ed. München 1965. In column 837 of this encyclopaedia, there is a drawing by Zonca. The design shown can hardly be regarded as credible.

1598 *Italy*

A more likely sketch of the same type of machine, from the same period and the same region, is included by Heinrich Schickhardt in his *Bericht über die Reise nach Italien 1598.* His manuscript is at the Württembergische Landesbibliothek, Stuttgart. The sketch has been published by Martin Eckoldt: *Der Brenta-Kanal bei Venedig in der deutschen Reise- und Fachliteratur des 16.–19. Jahrhunderts.* [Technikgeschichte, Vol. 35 (1968) fig. 4, p. 92]. See Fig. 40.

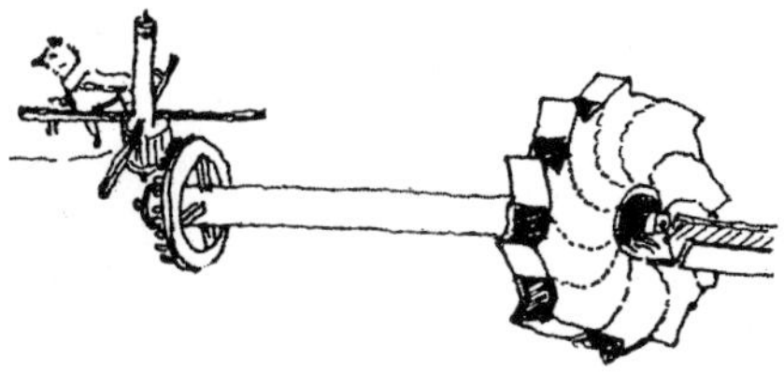

Fig. 40. Sāqiya with spiral scoop wheel in the Po Valley, Italy, 1598.

1598 *Persia*

Although the noria and the sāqiya are referred to as The Persian Wheel in Anglo-Saxon literature, there is no literature on the noria in Persia. Nor is there any mention of such machines, even in H. E. Wulff: *The Industrial Arts (Crafts) of Persia; their Development, Technology and Influence on Eastern and Western Civilisation.* Cambridge, Mass. 1966, although they were undoubtedly used in palaces and gardens and, to a lesser extent, in agriculture. Donald N. Wilber: *Persian Gardens and Garden Pavillions*, Tokyo 1962, p. 37 and p. 92:

> At the Ali Kapu Palace at Işfahān .. on the fourth floor ... water was raised by oxen power.
> .... three fountains once played in this basin, fed by water forced up by an ox-powered machine. About 1598?

There is an Arabic manuscript [*Oxford 954*] (cf. p. 63) mentioning water-raising machines from Iṣfahān, but none of these constructions resembles the noria or the sāqiya (cf. Vignette on p. 83).

I know the next work only from a German translation. Agostino Ramelli: *Le Diverse et Artificiose Machine* ... Paris 1588. A. Ramelli: *Schatzkammer der mechanischen Künste*, Leipzig, 1620, p. 199, fig. 87. This famous book covers a wide range of machines, but paternoster pumps are not mentioned, except for a single drawing of a treadwheel working a bucket chain.

1588
Paris

At the National Library in Madrid there is a unique manuscript containing a great deal of good engineering, but in the field of water-lifting wheels the designs tend to be rather unfortunate. Caro Baroja (1955), p. 62, presents facsimiles of two of these norias. A detailed discussion of this manuscript is available in Ladislao Reti: *The Codex of Juanelo Turriano (1500–1585)*. [Technology and Culture, Vol. 8 (1967), p. 53ff.]. The first drawing shows a high-lift sāqiya, but contains a number of errors; for example, the water pouring out of the buckets is hardly likely to hit the collecting trough. Again, however, it is not the mistakes we are interested in, but the fact that we find almost the same drawing later in Leupold and Zeising (cf. p. 51 and p. 52). The original drawing presumably came from Italy; at least, the same drawing is to be found in an Italian edition of Vitruvius dating from 1521 (cf. Vignette No. 8 on p. 151).

1500–1585
Madrid

Fig. 41. An Indian sāqiya irrigating a roof-garden. Detail from a miniature.

Let us turn for a moment from this somewhat abstract and plagiarizing European literature to Oriental art.

In the journal, Ars Orientalis, Vol. 3 (1959), p. 137, plate 4, we find an Indian miniature in which we can see a sāqiya located on the roof of one of the houses and can even distinguish the lantern pinion and the potgarland wheel, which raises water to the roof-garden. See Fig. 41.

16th cent.
India

There is another miniature in Josef Strzygowski: *Asiatische Miniaturenmalerei* ... Klagenfurt 1933, plate 40, fig. 119. Here, too, we have an urban scene, this time from Fatehpur in Northern India. The painting shows two tower-like buildings, at the top of which a couple of potgarland wheels can be glimpsed.

In European art, Caro Baroja has found an amusing picture of a high-lift sāqiya. This has been reproduced in the journal, Gazette des Beaux-arts 1892, p. 497, and is shown here in Fig. 42.

1502
France

The last authors of technical literature to be dealt with in this chapter are from Italy, the most famous among them being Leonardo da Vinci.

Th. Beck: *Leonardo da Vinci, Vierte Abhandlung*: *Codice Atlantico*. [Zeitschrift des Vereines deutscher Ingenieure, Vol. 50, (1906) p. 777]. This manuscript contains five water-raising contrivances without right-

1491
Italy

Fig. 42. Sāqiya of the elevated shaft type from France, about 1500.

angle gear. Four of them are pure perpetual motion machines but it is noteworthy that in one case, da Vinci attempts to prove that a water-wheel will consume more water than the machinery connected to it can raise. We shall not go into detail here regarding the errors of the demonstration, but merely note that da Vinci realized that there was, in fact, a problem and tried to solve it.

The rediscovered codices Madrid mss. 8937 and 8936 may have something more to tell us about water-raising contrivances, but they have not as yet (1972) been published. Ladislao Reti: *Die wiedergefundenen Leonar-*

do-Manuskripte der Biblioteca Nacional in Madrid. [Technikgeschichte, Vol. 34 (1967), pp. 193–225].

1475
Italy

Unfortunately, many manuscripts from the 15th century are still unpublished, including one by Martini. Ladislao Reti: *Francesco di Giorgio Martini's Treatise on Engineering and Its Plagiarists.* [Technology and Culture, Vol. 4 (1963), pp. 287–298]. This manuscript includes various water-raising machines, but these have not been published.

1438–1449
Italy

Pemp (cf. p. 149) has a manuscript illustration of an indoor sāqiya, which is attributed to Jacopo Mariano (Taccola). The design is correct, but there is no trough to collect the water from the bucket chain. More of the same author's work is available in J. H. Beck: *Mariano di Jacopo detto il Taccola. Liber Tertius de ingeneis ac edifitiis non usitatis*, edizioni il Polifilo. Milano 1969. This includes two plates showing geared norias that are driven by the wind.

Lynn White, Jr.: *Medieval Technology and Social Change*, Oxford U.P. 1962, p. 131, includes the very realistic indoor sāqiya mentioned above in conjunction with perpetual-motion machines. Professor White finds points of similarity between one of these machines and a machine dealt with in my next chapter under the title, *Wheels which turn by themselves*. Professor White's book, which is a basic work on the history of technology, will be cited frequently in the following.

Vignette No. 3. Shiraz painting, Iran, A.D. 1341 Fogg Art Museum, Cambridge, Mass.

CHAPTER 3

Islamic Waterwheels and Perpetual Motion

Literature from A.D. 700 to A.D. 1400 reviewed in inverse chronological order

Arab literature

Arab literature contains many references to water-raising devices, but only a few of these relate to the geared noria. In the following, the wheels at Ḥamā will also be discussed, although they actually lie outside the scope of this study.

1362 *Ḥamā*

The largest wheel bears the name nāʿūra al Muḥammadiyya. It must date from the 14th century as an inscription on the aqueduct reads:

> This big and blessed water-wheel was constructed . . . in the days of our Lord, His Most Noble Excellency Saif ad-dīn Ṭānyaraq, Viceroy of the province of Hama, during the last days of the year 763 [i.e. October 1362.]

This inscription can be found in L. A. Mayer: *Saracenic Heraldry*, Oxford 1933, p. 224.

E. Wiedemann

The subject is a very involved one, but, fortunately, E. Wiedemann has extracted the passages dealing with technical (physical and chemical) matters. It is something of a drawback that E. Wiedemann makes great demands on his readers; it does not suffice to know the German language – readers are also required to distinguish between translated texts and E. Wiedemann's own comments, since no typographical difference is made between them. The references are incomplete and it is impossible to find them without a bibliography. A number of the references have proved to be original dictionaries dating from the fourteenth century. E. Wiedemann: *Beiträge zur Geschichte der Naturwissenschaften III.* [Sitzungsberichte der Physikalisch-medizinischen Societät in Erlangen, Vol. 37 (Erlangen 1905 (1906)), pp. 218–263]. This and subsequent articles with the same title and from the same journal will be denoted in the following as: [Beiträge *III*, Vol. 37], [Beiträge *V*, Vol. 37, pp. 392–445], [Beiträge *VI*, Vol. 38, pp. 1–56], [Beiträge *X*, Vol. 38, pp. 307–357].

Dimašqī's work, which has been translated by A. F. Mehren: *Manuel de la cosmographie du moyen âge* ... Copenhague 1874, p. 281, see also [Beiträge *VI*, Vol. 38, p. 51], reads:

1327
Dimašqī

> Along the Orontes [river in Syria] there are large norias, the like of which are not seen anywhere else on earth. They supply water [to the canals] from the Orontes, which in turn irrigate the gardens and squares.

The machine at Ṣafad (Northern Israel) described by Dimašqī is not a sāqiya, even though it is driven by a mule. See Guy Le Strange: *Palestine under the Moslems*. London 1890, pp. 524–525.

1251
Taʿāsīf

An Arab engineer, known by the name of Taʿāsīf (dec. 1251), is cited by *Abulfedae annales muslemici, arab. et lat. opera et studiis* by J. J. Reiskii *et al.* ed. J. G. Adler; Hafnia (Copenhagen) 1789–1794. Vol. 4, p. 479. See also [Beiträge *VI*, Vol. 38, p. 44]:

> He built several towers in Ḥamā and a mill (noria?) by the Orontes. He was an able mathematician.

The last remark should presumably be understood to mean: He was a good geometrician and craftsman.

1206
Ǧazarī

A very important contribution to our understanding of Arab technology is provided by Ǧazarī, but this is so extensive that it will be dealt with in a separate section. This also applies to another contribution, by ibn al ʿAwwām.

12th century
Cairo

The Metropolitan Museum, Egyptian Expedition: *The Monastery of the Wādi 'n Natrūn*. Part *II*, N.Y. 1932 ed. W. Hauser p. 382:

> These objects, he stated, were gifts from Christians, and had the donors' names inscribed upon them. They were then produced and taken to Cairo, where they were shown to El Kāmil. Among them was a waterwheel of crystal of wonderful workmanship and a network of pearls.

I thought at first that the waterwheel was a model or toy, but have since realized that the objects in question must be jewels, in view of the fact that Lane has a drawing of a gold filigree charm, circular in shape and called a sāqiya. See E. W. Lane: *An Account of the Manners and Customs of the Modern Egyptians* ... Vol. *II*. London 1836, p. 358.

1184
ibn Ǧubayr

The Travels of Ibn Jubayr, translated by R. J. C. Broadhurst, London 1952, pp. 252–253 and 265:

> An account of the city of Ḥamā. May God Most High defend it. ... To the east of the city I saw a large river that in its strong course spreads out and branches, and on its banks observed water-

wheels that faced each other [two wheels sharing a single shaft]. Along these banks are disposed gardens that hang their branches over the water, the green leaves appearing like down on its cheeks as it flows through their shade, gliding through the line of their symmetry.

An account of the city of Ra's al ʿayn. May God protect it. (Halfway between Mosul and Aleppo).

. . . it has a garden and beside it is a waterwheel that sends the water to a garden that is higher than the level of the river. . . . . If the villages of eastern Andalusia had such a beautiful site or such sweet springs they would be of the very best. God disposes of all that He creates.

1066
ibn Sīda

There is a description of the sāqiya in a 17-volume Arabic encyclopaedia written by ibn Sīda, who died in 1066. This description is given by E. Wiedemann und F. Hauser: *Über Vorrichtungen zum Heben von Wasser in der islamischen Welt.* [Beiträge zur Geschichte der Technik und Industrie, Vol. 8 (Berlin 1918) p. 129]. The German translation is given below:

The dawlāb.
On top of the wheel there are two strong endless ropes. Between these there are pots and they look like small, pitched palm baskets but are called ʿuṣmūr.
The length of the ropes depends on the distance between the water and the place where the pots are emptied. As the wheel turns, the pots rise and on the other side they descend. In this way the empty pots take water and the filled pots rise. At the top they turn round and pour their liquid into a channel of wood. By the channel is the machine (manganūn), which is turned by the camel, the ox, or the donkey.

Das *Daulāb* oder *Dūlāb* (Wasserrad) . . . . .
Auf seinem Rücken befinden sich zwei starke Stricke (*Masad*). Die Enden eines jeden sind miteinander verbunden. Zwischen ihnen werden Krüge (*Kūz*) angebracht, die kleinen Eimern (*Dalw*) aus Palmfiedern gleichen; sie sind verpicht. Diese Krüge heißen *ʿUṣmūr* (auch *Qādūs*). Die Stricke haben eine Länge, die dem Abstand zwischen dem Wasser und dem Ort, wo die Eimer ausgegossen werden, entspricht. Dreht sich das Rad, so steigen die Eimer auf der einen Seite, und die diesen gegenüberliegenden sinken auf der anderen Seite. So schöpfen die leeren Eimer Wasser, und die vollen steigen in die Höhe. Haben sie die Höhe des Rückens des *Schahraq* erreicht und wollen eben umkehren, so gießen sie ihren Inhalt in einen Kanal aus Holz. Über dem Kanal steht aber der *Manganūn* (Maschine); diese Maschine dreht das Kamel, das Rind, der Esel.

In a frequently used bibliographical work, *Kitāb al Fihrist*, which can be traced back to 987, and which was published by Gustav Flügel in Leipzig in 1871–1872, we read that:

987
al Fihrist

> Harqal the carpenter has made a book about the wheels and the waterwheels (dawlāb).

There is no means of telling whether this work is identical with *Wheels which turn by themselves* (cf. p. 72), but Wiedemann suggests that this may be the case [Beiträge *VI*, Vol. 38, p. 4].

There are also Arabic papyri, in one of which the terminology is the same as that used at the present time.
P. Groh. [Archiv Orientální, Vol. *XVIII* (1950), pp. 90–92]:

950
Papyrus

> .... three sāqiyas, rent 5 dinars.

In a work by al Balāḏurī (dec. 892), published in Arabic by W. Ahlwardt, Leipzig 1883, p. 370, we read:

892
Balāḏurī

> In Baṣra, on one of the farms founded by Muḥammad ibn Sulaymān, there were water tanks (ḥawḍ), which were filled by camel-driven waterwheels (dawlāb). [.... die durch von Kamelen in Bewegung gesetzte Wasserräder gespeist wurden.]

[Beiträge *X*, Vol. 38, p. 333]. There is an English translation of this passage, which mentions "the waterwheels and the camels", but not "camel-driven waterwheels". See Francis Clark Murgotten: *The Origins of the Islamic State ..... al-Balādhurī.* (Columbia University) N.Y. 1924, Vol. 2, p. 98.

Jacut's geographisches Wörterbuch aus den Handschriften zu Berlin, St. Petersburg und Paris .... ed. Ferdinand Wüstenfeld. Leipzig 1867, Vol. 2, p. 331, in Arabic. Yāqūt writes:

1229
Yāqūt

> On the Orontes there are several waterwheels, which raise the water so that the garden can be irrigated and pour water out in the cistern of the mosque.

[Beiträge *VI*, Vol. 38, p. 51]. There is, in fact, a mosque at Ḥamā at which there is a big wheel, and I presume that this is the one mentioned by Yāqūt. The wheel, which stands outside the mosque, is still in operation (1969).

Yāqūt found an even older reference in Aḥmad ibn al Ṭayyib, who died in 885:

885
Ṭayyib

> Ḥamā is a town with a stone wall, beside which there is a big stone building. The river flows and waters the gardens and drives their

waterwheels (nāʿūra). This report comes from the year A.H. 271 (A.D. 885), and he calls his account "A town".

This is the oldest text referring to the great wheels at Ḥamā or, rather, the oldest text found by Wiedemann. There are a number of even older Arabic texts mentioning the Egyptian sāqiya.

850 *Papyrus*

P. Groh. Adolph Grohmann: *Arabic Papyri in the Egyptian Library*, Vol. *II* Cairo 1936, pp. 42–43:

> .... Lease of 3 faddān ... irrigated by the sāqiya (min arḍ as-sāqiya) to the south of the canal ....

830–849 *Alexandria*

History of the Patriarchs of the Coptic Church of Alexandria, No. IV, ed. and translated by B. Evetts. [Patrologia Orientalis, Vol. *X* (Paris 1915) p. 516]:

> Listen now to yet another wonder, my friends, concerning this father (abba Joseph 830–849). There was at Alexandria a man who was one of the Chalcedonian heretics; and he was very rich, and possessed water-wheels. So when he went out one day to take recreation in one of his vineyards outside Alexandria, he saw a broken water-wheel, and asked for a carpenter. Now there was an old man, a priest, who was also a carpenter. So the Chalcedonian said to him: "Wilt thou come now with me, that thou mayest mend the water-wheel for me?" But the priest replied: "Today is the Great Friday; and I can do no work today, because it is the day on which the Word of God, the Saviour of the world, was crucified." Then that cursed heretic opened his cavilling mouth in answer, and blasphemed God the Word, saying what must not be recorded. So the aged priest reproved him, and went away and left him."

The text is old, but the characters live on. There are many rich people in Alexandria who have sāqiya-watered vineyards outside the town and who still use them for recreation on holy days.

762 *al Kindī*

I found the previous and the next contribution in Dieter Müller-Wodarg: *Die Landwirtschaft Ägyptens in der frühen ʿAbbāsidenzeit 750–969, n. Chr. (132–358 d. H.)*. [Der Islam, Vol. 31 (Berlin 1953), pp. 174–227]. A work by al Kindī mentions Yazīd ibn Ḥātim (dec. 762), who had a sāqiya built to feed a cistern. Rhuvon Guest (editor): *The Governors and Judges of Egypt*. Leiden 1912; in E. J. W. Gibb Memorial Series, Vol. *XIX*, p. 115 in Arabic.

750 *Marwān*

Müller-Wodarg has unearthed yet another reference in: *History of the Patriarchs of the Coptic Church of Alexandria*; ed. and translated by B. Evetts. [Patrologia Orientalis, Vol. *V* (Paris 1910), p. 169]:

> Before the last ʿUmayyād-caliph Marwān *II* fled to Upper Egypt,

he ordered his soldiers to set fire to towns, villages, vineyards and sāqiyas north (of the town) of al Fusṭāṭ.

In the following sections, a number of manuscripts of purely technical content will be discussed. The first of these is an Arabic translation of Philon's *Pneumatics*, in which several waterwheels are mentioned.

Wiedemann has found the following reference in a thesis by J. Østrup: *Umar ibn Muhammed al-Kindis Beskrivelse af Ægypten*, resumé en français. Oversigt over det Kongelige Danske Videnskabernes Selskabs Forhandlinger, Nr. 4 København 1896, p. 173–245. [Beiträge *III*, Vol. 37, p. 230]:

About 970
ʿUmar al Kindī

Al Kindī says of Philon that:
... he was well-versed (in the design) of waterwheels (dawlāb) and mills, and in movements effected by ingenious devices (ḥiyal).

Philon is also mentioned by al Yaʿqūbī, who is cited in: *Ibn-Wādhih qui dicitur al-Jaʿqubī Historiae*; ed. M. Th. Houtsma, Lugduni Batavorum (Leiden) 1883, p. 135, line 4, in Arabic:

891
al Yaʿqūbī

Philon is one of the inventors of maḫānīqā (pneumatics).

The next few sentences are completely incomprehensible, but concern waterwheels. They are followed by the interesting sentence:

Devices that are driven by water without being driven (by man or beast).

Here we encounter for the first time the attitude of Islamic scholarship to the concept of perpetual motion; they regarded any type of machine not powered by man or beast as a perpetual-motion machine [Beiträge *V*, Vol. 37, p. 438].
The Arabic manuscripts of purely technical content usually cover a wide range of subjects that we would not regard as having anything at all in common. The Arabs, however, considered them to be related subjects and thought they should be collected in a single book.

Al Anṣārī (dec. 1348) is cited by Ḥāğī Ḫalfa (dec. 1657), whose work was translated into Latin by Gustavus Flügel: *Lexicon bibliographicum* ... Leipzig-London 1835–1858. [Beiträge *V*, Vol. 37, p. 396]:

1348
al Anṣārī

From (geometry) ten sciences can be derived: architecture, optics, burning mirrors, centres of gravity, measuring and weighing, *lifting water*, lifting loads, clocks, war-machines and *pneumatic apparatuses.*

To these might well be added two others, viz, horsemanship and Greek fire.
We will now move on to another category of material, which deserves study in greater depth.

Illustrated Arabic manuscripts

Introduction

In the following, four works on water-raising machinery will be discussed. The first is by the Greek student of mechanics, Philon of Byzantium, whose *Pneumatics* includes a number of waterwheels. As we know his work only through Arabic translations, it is included in this section, even though Philon actually lived in the 3rd century B.C.

The next work, which is due to Ğazarī, appears to be purely Arabic, but a detailed study shows clear indications of Indian and Greek influence. The manuscript, of which there are about 15 different copies, is famous for its rich illuminations, but its technical content remains almost unknown. It is, in fact, one of the most important contributions to mediaeval technology. The best – but as yet unpublished – manuscripts are in Istanbul, and the poorest are to be found in the Western hemisphere.

Another manuscript of which there are also several copies bears the title, *Wheels which turn by themselves*, and, as indicated by the name, discusses perpetual motion. The designs are of no practical significance, but they are interesting in that they represent the attitude of that period to problems of this kind.

As a down-to-earth contrast to the realm of fantasy, there is fortunately an agricultural manual by ibn al ʿAwwām, which describes suitable methods for improving the noria.

List of manuscripts on water-raising machines

Manchester 351

Location: John Rylands Library, Manchester.

Catalogue entry: Arabic MS. cod. 351 (Hydraulic machines, folios 15b to 50b, 52b, 53a and 84a). See also A. Mingana: *Catalogue of the Arabic Manuscripts in the John Rylands Library, Manchester*. Manchester U.P. 1934, columns 552–560.

Containing: Heron, Philon, Apollonius and various anonymous authors.

A single illustration, possibly depicting a waterwheel, is to be found in: H. J. J. Winter: *Muslim Mechanics and Mechanical Appliances*. [Endeavour, Vol. *XV* (London 1956), p. 26, fig. 1].

Istanbul 2755

Location: Süleymaniye U. Kütüphanesi, Istanbul.

Catalogue entry: Hagia Sophia 2755. (Hydraulic machines, folios 61v to 69v) (Philon's water lifting machine, folio 145v).

The drawings, which are done in red and black Indian ink, have not previously been published. See Plate 1.

Istanbul
3713

Location: See Istanbul 2755.

Catalogue entry: Hagia Sophia 3713. (Philon's water-lifting machine, folio 84r). A French translation is to be found in: Carra de Vaux: *Le livre des appareils pneumatiques et des machines hydrauliques, par Philon de Byzance* ... [Notices et extraits des manuscrits de la Bibliothèque nationale et autres bibliothèques, Vol. 38 (Paris 1903), p. 27 ff]. Carra de Vaux's interpretation has been translated into German by Th. Beck: *Philon von Byzanz.* [Beiträge zur Geschichte der Technik und Industrie, Vol. 2 (Berlin 1910), pp. 64–77].

The drawing in Fig. 45 is done in red and black Indian ink.

Oxford
954

Location: Bodleian Library, Oxford.

Catalogue entry: Oxford 954 Marsh 669 (Hydraulic machines between folios 3v and 9v). A classification of the contents of the manuscript is available in: Carra de Vaux: *Notice sur un manuscrit arabe traitant de machines attribuées à Héron, Philon et Archimède.* [Bibliotheca mathematica, 3rd Ser. Vol. 1 (Leipzig 1900), pp. 28–38].

The drawings in Plate 2 have all been published previously, although not together.

The water-lifting contrivances from "Die Leute von Iṣfahān" mentioned by Carra de Vaux (1903) and Wiedemann und Hauser (1918), pp. 149–154, are contained in this m.s., folios 13v–17r. Hauser's copies are very accurate, lacking only an outlet in square h at the point marked z in fig. 36.

Istanbul
3606

Location: See Istanbul 2755.

Catalogue entry: Hagia Sophia 3606. This is apparently a complete Ğazarī-manuscript, in marvellous colours. Carra de Vaux (1900), p. 28, has commented on it as follows:

> It is illustrated with innumerable figures coloured with great care in a Persian and Indian style. This is one of the most beautiful Arabic manuscripts I have ever seen.

I myself have had the great pleasure of inspecting the manuscript. It is not only of singular beauty but must also be regarded as the best manuscript known to us. I have arrived at this conclusion after comparing the same motif of a machine in six different manuscripts; folio 222v of that under discussion is the only one completely free of technical errors.

Istanbul
3472

Location: Topkapu Serai Müzesi Kütüphanesi, Istanbul.

Catalogue entry: T. K. S. Müzesi A 3472.

This Ğazarī-manuscript can be dated to 1207 (A. H. 602), not 1254, and must therefore be regarded as the oldest known Ğazarī-manuscript. It comprises 179 folios with 60 illustrations. Some of the illustrations are given in: I. Stchoukine: *Un manuscrit du traité d'al-Jazarī sur les automates du VII s. H.* [Gazette des beaux-arts, Vol. 6, (1934), pp. 134–140]. Folio 136v is to be found in colour in: Ekrem Akurgal, Cyril Mango and Richard Ettinghausen: *Treasures of Turkey*. Geneva 1966, p. 170.

Istanbul 414

Location: See Istanbul 3472.

Catalogue entry: T. K. S. Müzesi H 414.

This manuscript is not listed in any of the catalogues, even though it must be regarded as the second-best Ğazarī-manuscript. It contains 174 folios and 122 coloured illustrations. There is a microfilm of the manuscript in the League of Arab States, Manuscript Department, Cairo. See Plate 3.

Istanbul 3350

Location: See Istanbul 3472.

Catalogue entry: T. K. S. Müzesi A 3350.

This is one of the earlier copies, from 1458. It has 253 folios and 144 rather poor illustrations.

Istanbul 3461

Location: See Istanbul 3472.

Catalogue entry: T. K. S. Müzesi A 3461.

Despite its quality, I have found no references to this manuscript. The illustrations on folios 189v and 190v correspond to the motif shown in Figs. 47a and 47b. Folio 192v corresponds to Fig. 48. The manuscript comprises a total of 211 folios and 143 illustrations.

Oxford 27

Location: See Oxford 954.

Catalogue entry: Oxford Greaves (not Graveley) 27.

This is the only Ğazarī-manuscript that has been translated. Unfortunately, the translations are not in a combined work, but they can be found by means of J. D. Pearson's catalogue: *Index Islamicus 1906–55*, Cambridge 1958, and *Index Islamicus, Supplement 1956–60*, Cambridge 1962.

A translation of the part on water-lifting contrivances is given in: E. Wiedemann (1918), pp. 137–147.

Paris 2477

Location: Bibliothèque nationale, Paris.

Catalogue entry: Fonds Arabe 2477, from the year 1485.

This Ğazarī-manuscript has 112 folios and 31 illustrations. There are no illustrations to the chapter on water-lifting machines.

Paris 5101

Location: See Paris 2477.

Catalogue entry: Fonds Arabe 5101, from the 18th century. A total of 253 folios but no illustrations.

Wheels which turn by themselves

Leiden 499

Location: Biblioteek der Rijksuniversiteit, Leiden.

Catalogue entry: Codex Orientalis 499, or CCO 1414 (Hydraulic machines, folios between 74v and 87r).

Istanbul 1884

Location: See Istanbul 2755.

Catalogue entry: Esad Efendi 1884 or 'Asʿad 1884 (Hydraulic machines, folios between 71r and 85v. See Plate 4).

Gotha 1348

Location: Universitäts-Bibliothek, Gotha.

Catalogue entry: Codex Gotha 1348 (Hydraulic machines, folios between 109v and 124r). [Beiträge *III*, Vol. 37, p. 231] and [Beiträge *VI*, Vol. 38, p. 13].

Istanbul 3469

Location: See Istanbul 3472.

Catalogue entry: K. T. S. Müzesi A 3469 (Hydraulic machines between folios 127r and 184r).

H. Ritter: *La Parure des Cavaliers und die Literatur über die ritterlichen Künste*. [Der Islam, Vol. *XVIII* (Berlin 1929), p. 152].
This magnificent manuscript has 235 folios and 108 illustrations depicting catapults, scaling ladders, a catapult with elevation scale, crossbow, and Greek fire. The illustrations to the chapter on water-lifting machines are almost the same as those on Plate 4.

As mentioned earlier, Oxford 954 and Istanbul 2755 also include the chapter on *Wheels which turn by themselves*. (i.e. Hydraulic machines).

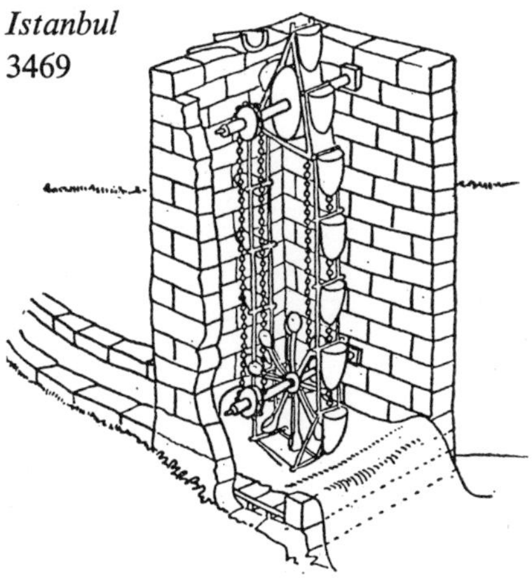

Fig. 43. Reconstruction of Philon's water-lifting machine. The construction has no counterpart in practice and it is doubtful whether it can be attributed to Philon, who lived about B.C. 300.

Philon of Byzantium

Almost all the books on the history of technology include a copy of Carra de Vaux's reconstruction of Philon's water-lifting machine. I have included it as well (Fig. 43), but in Caro Baroja's own version. Julio Caro

Fig. 46 and Plate 3. Title-page of Ǧazarī manuscript. Al Ǧāmiʿ bain al ʿilm wal ʿamal an nāfiʿ fī ṣināʿat al ḥiyal. In the author's translation. "All between theory and work applied to construction of mechanical magic". Other copies of the same manuscript are entitled: Kitāb fī maʿarifat al ḥiyal al handasiyya. "Book of knowledge of geometrical, i.e. mechanical contrivances". Redrawn by the author from TKS Müzesi H 414.

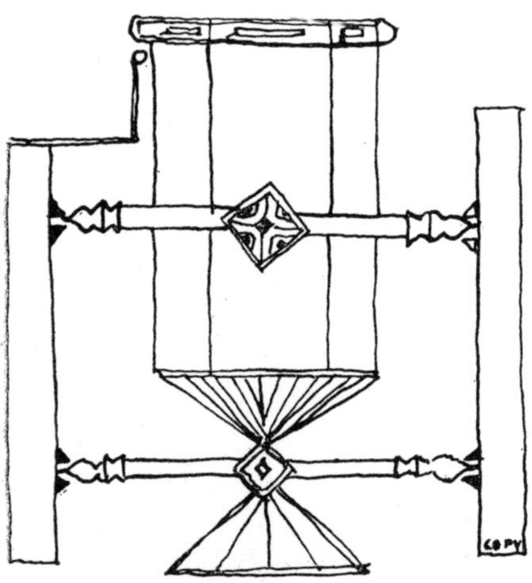

Fig. 44. Manuscript drawing for Philon's water-lifting machine. The drawing is not completely comprehensible. Hagia Sophia 2755. Redrawn by the author.

Baroja: *Sobre maquinaria de tradición antigua y medieval.* [Revista de dialectología y tradiciones populares, Vol. 12 (Madrid 1956), p. 132, fig. 18]. It is impossible to understand either the text or the illustrations (Figs. 44 and 45), and Carra de Vaux's reconstruction is questionable. Dr. Drachmann has told me that the propelling mechanism is at any rate incorrect. According to the text, the machine is supposed to be a waterwheel driven by falling water.

What is strange is the vertical chain drive. It is a golden rule that the power should be applied to the upper wheel. In this case, the waterwheel turns the lower one. Furthermore, the chain drive ought to be slanting.

The crucial question is not whether the design is serviceable, but whether it can really be attributed to Philon; it seems more likely that it derives from a later inventor. Although the Arabic letters used to label the figure follow the order of the Greek alphabet, this does not prove that the manuscript is a translation from the Greek. This problem has been dealt with by S. Gandz: *Der Hultsch-Cantorsche Beweis von der Reihenfolge der Buchstaben in den mathematischen Figuren der Griechen und Araber.* [Quellen und Studien zur Geschichte der Mathematik, Astronomie und Physik, Abt. B, Studien 2, 1932 (1933), pp. 81–97].

The literature on Ǧazarī is extensive, but most of it concentrates only on the artistic aspect of the illustrations. There is a Ǧazarī-bibliography in: Kurt Holter: *Die islamischen Miniaturhandschriften vor 1350.* [Zentralblatt für Bibliothekswesen Vol. 54 (1937) pp. 5–8].

H. Buchthal, O. Kurz and R. Ettinghausen: *Supplementary notes to K. Holter's check list of Islamic illuminated manuscripts before A.D. 1350.* [Ars Islamica, Vol. *VII*, (Ann Arbor 1940) pp. 148–149].

Ǧazarī's work comprises six chapters: 1) Design of water clocks; 2) Design of casks suitable for wine-parties; 3) Design of containers for blood-letting and washing; 4) Design of balances with containers and an instrument that always whistles; 5) On the design of instruments that raise water from a large mass, and a well that is not deep, and from a flowing river. (This is the only chapter heading that is a literal translation of the German version [Beiträge *III*, Vol. 37, p. 261]); 6) Design of various figures. For the sake of completeness the original German text is given below:

> 1. Über die Konstruktion der Wasseruhren, durch die man den Ablauf der gleichmäßigen und zeitlichen Stunden kennen lernt (in 10 Propositionen). 2. Über die Konstruktion der Gefäße und Formen, die zu Weingelagen geeignet sind (in 10 Propositionen). 3. Über die Konstruktion der Krüge und Becher (Becken (ṭisās)) zum Aderlassen und zur Waschung (in 10 Propositionen). 4. Über die Konstruktion der Flaschen (Gefäße) auf einem Knie, die ihr Lage wechseln, und das Instrument der ewigen (nie aufhörenden) Flöte (in 10 Propositionen). 5. Über die Konstruktion der Instrumente, welche das Wasser heben aus einer großen Masse und einem Brunnen, der nicht tief ist, und aus einem fließenden Flusse (in 5 Propositionen). 6. Über die Konstruktion verschiedener Figuren, die ihm nicht ähnlich sind (in 5 Propositionen).

As will be seen from the headings, the work concerns *wonderful designs*, a concept that the Arabs designated ḥiyal.

Although Ǧazarī's machines are toys, they do tell us something about the mechanical thinking of the period.

The First Figure, Fig. 47a, shows an improved version of a design known in India. There, a big scoop made from a hollowed tree-trunk is used. One end of the trunk rests on the bank, while the other floats on the water. Over the trunk, a šādūf-type structure is built. The vertical šādūf-rope is fastened to the outer end of the tree-trunk. When the peasant walks out along the trunk, this fills with water; when he returns, the counterweight of the šādūf lifts the tree-trunk, and the water gushes out onto the field. The method is only suitable for very small lifts.

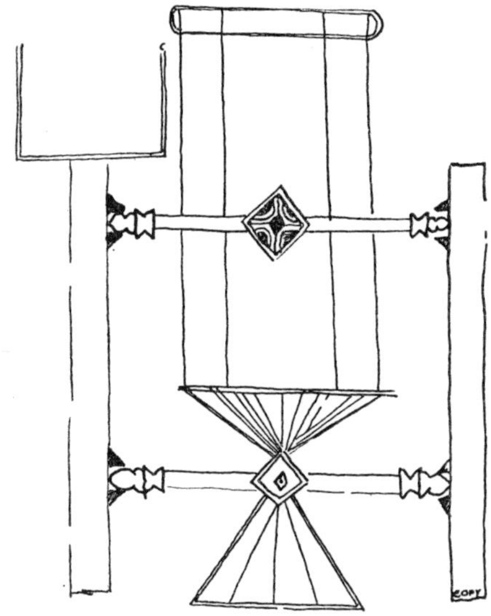

Fig. 45. Manuscript drawing for Philon's water-lifting machine. Hagia Sophia 3713 folio 84 recto. Redrawn by the author.

Fig. 47a and Plate 3. Ǧazarī, TKS Müzesi H 414, folio 151 verso. See the text to the next figure on p. 68.

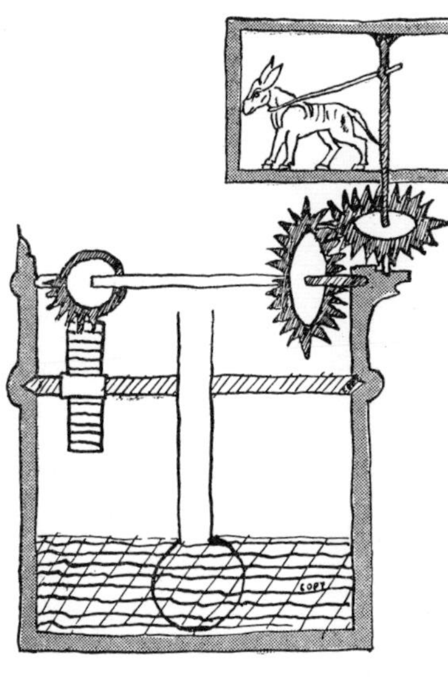

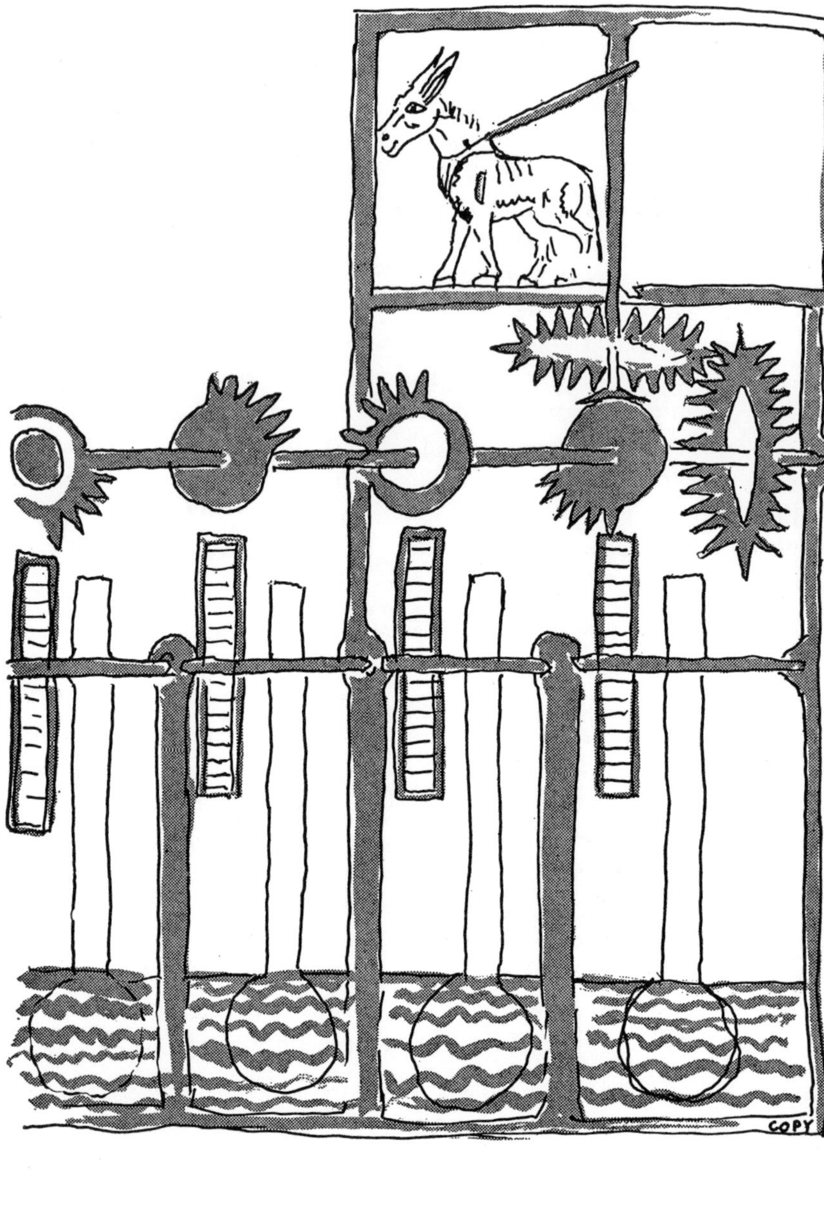

Fig. 47b and Plate 3. Ğazarī, TKS Müzesi H 414, folio 152 recto. An ass is turning a right-angle gear, which in turn drives four cog-wheels with radial cogs, which mesh with four lantern pinions. The wheels with the radial cogs have cogs over only a quarter of their circumferences, which means that the lantern pinions can – one by one – turn 90°. Then the scoops are lifted. These must be designed as bailers on the ends of long tubes. In their vertical position, the bailers fills up with water, and in their horizontal position, the water runs out. Ğazarī stresses that the cogs on the four uppermost cog-wheels must have a phase shift of four times 90°.

Fig. 48 and Plate 3. Ğazarī. TKS Müzesi H 414, folio 153 verso. An impact wheel turns a right-angle gear, which in turn makes a wooden ox move round. At the same time, the top right-angle gear turns a bucket-chain wheel.

Ğazarī replaces the peasant with an ass. A special cog-wheel with teeth over one-quarter of its periphery meshes with another cog-wheel, which lifts the scoop. When the teeth unmesh, the scoop falls back to its vertical idle position and fills up again. It appears from the text that the invention was intended as a toy. The structure has no counterpart in fact.

The Second Figure, Fig. 47 b, shows the same design, but here provided with four scoops.

The Third Figure, Fig. 48, shows a sāqiya with an elevated shaft. This, too, is intended as a toy, but is of the same size as a real paternoster pump. Hidden under the watercourse is an impact wheel, which drives a wooden ox round by means of a right-angle gear. To the spectator, the wooden ox appears to be turning the wheel.

This is the only *Figure* that is closely related to our subject, so an

English translation, possibly with an Arabic-German flavour, is given here for the first time.

Wiedemann and Hauser, p. 141, figs. 16 and 17. See Fig. 49.

> *The Third Figure.* At the centre of a pool (birka), there is a column, on which a disc is placed. On this a model of an ox stands turning a wheel; the device lifts the water about 10 *spans* above the pool. Unlike the water-lifting devices already described, this one is not driven by animal power, but by water power. [10 *spans* is 2.5 m].
> Into a rectangular pool (s) with a sheet-copper bottom and marble sides, water flows from the right-hand side, running through a pipe to a brook or similar. About two-thirds of this water falls through a pipe (e), which is fixed to the bottom of the pool, and hits a shell-wheel (h) located beneath the pool, thereafter flowing out through a channel at the right-hand side. All, or part, of the last third of the incoming water is supposed to be raised a certain height. Any surplus water passes out through the pipe shown on the left-hand side. The shell-wheel (h) should have large shells so that most of the mass of the falling water can be used for work. The shell-wheel is placed on a 3-*span* long horizontal shaft and turns the upright shaft by means of two gear-wheels. The upright shaft (j) is made of iron and is slender. It reaches a height of 8 *spans* above the pool.

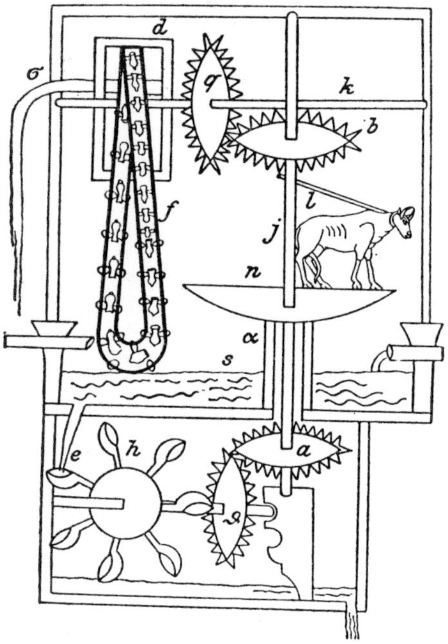

Fig. 49 above and below. Ğazarī. Copy of Oxford Greaves 27, with Wiedemann's notation and measurements.

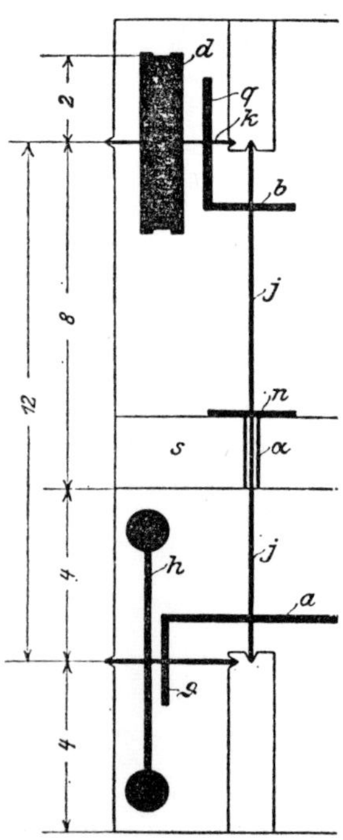

The next sentence is very important because it concerns a concept encountered in Indian literature.

> The means of propulsion is kept hidden (a secret) from the spectator, in an 8-*span* deep chamber under the pool. To him it appears as though the movement is achieved by an ox pulling a pole (l).

It can hardly be fortuitous that the same idea – with the same secret source of power – is to be found in an Indian manuscript; V. Raghavan: *Yantras or Mechanical Contrivances in Ancient India.* (The Indian Institute of Culture, Basavangudi, Bangalore, Transaction No. 10, 2nd ed. (India 1956), p. 22). The manuscript, which is in the Jesalmere Bhandars, reads as follows:

> ... the machines are then divided into two classes, Svayam-vāhaka, automatic, and Sakrit-prerya, requiring occasional propelling; most machines combine these two features. Another classification is: the concealed, Antarita or Alakshya, *i.e.* the principle of its action and its motor-mechanism are hidden from public view.

In this manuscript of Bhoja (1018–1053) mention is made of both drinking elephants and, further on, apes and peacocks, and it is precisely these animals that we encounter in the manuscript of Ğazarī.

Dr. Joseph Needham has discovered other points of similarity between

Indian texts and Ğazarī [Needham, Vol. 4, Part *II*, p. 539, note j]. Kurt Weitzmann, too, is interested in finding Ğazarī's source. See Kurt Weitzmann: *The Greek Sources of Islamic Scientific Illustrations*. (Archaeologica Orientalia in Memoriam Ernst Herzfeld, New York 1952, p. 249).

After this slight digression we now continue with the translation of Ğazarī's *Third Figure*:

> It is not, however, a live ox, but only a small, presumably lightweight, hollow, wooden model, a beautifully painted thing. The model is fastened by means of a rope around its neck to the pole (l) of the shaft (j) (sahm madār).
>
> The model is carried by the pole revolving with it. In this way the model is moved around barely above the horizontal copper disc (n) that serves as the circulat track, but without actually touching this.

The next sentence is not quite clear, but its meaning must be as follows:

> The disc (n) is placed halfway up the upright shaft (j) and rests on the hollow copper column (α) surrounding the shaft and fastened at the middle of the bottom of the pool, over and concentric with the hole holding the shaft and at the same level as the edge of the pool.
>
> Above the ox, the shaft (j) has a geared wheel (b), by means of which it works the geared wheel (q) and its shaft (k). On this shaft, a 4-*span* diameter wheel (d) is placed. Over this run two endless silk ropes with a row of slender copper buckets fastened between them. The buckets, which have the same width at top and bottom, each contain about 30 *dirhams* (90 grams). The buckets are fastened to the ropes by means of two arms (plugs) at top and bottom. These arms stand opposite each other (*i.e.* they are parallel).
>
> The ropes and buckets dip into the water, almost to the bottom of the pool (s), and the buckets lift the water to the top (of the wheel), where they pour it into a chute (σ), from which it runs out for further use.

The potgarland wheel is called a sindī, which implies that it comes from the Province of Sind in Pakistan.

The Fourth Figure, Fig. 50, also shows an ox-driven machine. This is yet another water-scooping device that has to be moved up and down. To achieve this, Ğazarī uses a crank movement. The drawing is not very easy to follow because all the parts are turned into the plane of the drawing. However, if we imagine the horizontal shaft placed so that it stands out of the plane of the drawing, the pin of the crank will fit into the scoop slot. As the crank turns, the scoop rocks up and down.

The Fifth Figure, and the last in this chapter on water-raising machines, deals with a double-acting piston pump.

Fig. 50 and Plate 3. Ğazarī. TKS Müzesi H 414, folio 155 recto. An ox turns a right-angle gear, which in turn drives a crank. In order to understand the next part it is necessary to imagine the horizontal shaft with the crank turned in such a way that the shaft stands perpendicular to the plane of the drawing. The crank will then fit into the slot in the diagonal tube that ends in a bailer. As the crank turns, the scoop moves up and down.

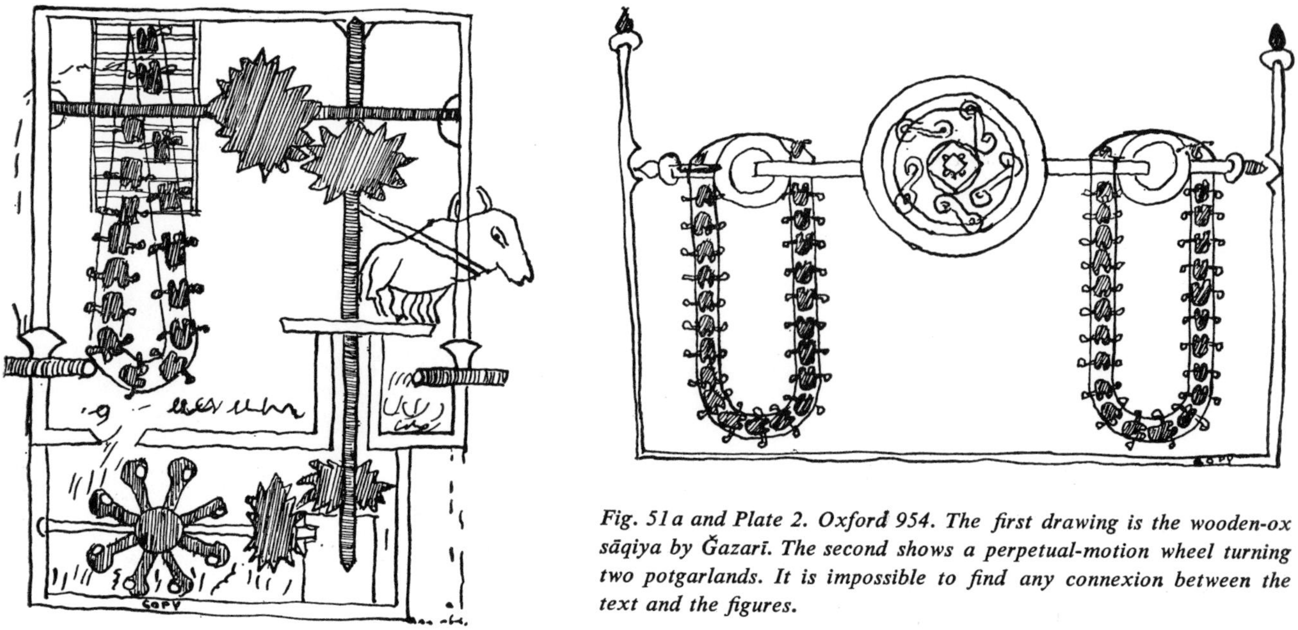

Fig. 51a and Plate 2. Oxford 954. The first drawing is the wooden-ox sāqiya by Ğazarī. The second shows a perpetual-motion wheel turning two potgarlands. It is impossible to find any connexion between the text and the figures.

Wiedemann's article omits the name of the manuscript used, but as mentioned earlier, this is Oxford Greaves 27. The drawings have been made with great care by Hauser. I have compared them with the drawings in T. K. S. Müzesi H 414, and have found a total of 17 errors or defects of a technical nature. I used the same procedure in the library at Istanbul and found in this way that Codex Hagia Sophia 3606 has the best drawings and can perhaps be attributed to the master himself. It is in any case obvious that the draughtsman understood the designs.

It should be emphasized that Ğazarī's book is a book on automata, not a treatise on practical water-raising machines. It is astonishing to find that the designs are completely free of the concept of perpetual motion. This is by no means true of the next manuscript, which deals with water-raising machines driven by perpetual-motion machines.

Wheels which turn by themselves

This chapter heading is to be found in six different manuscripts (cf. p. 65).

Carra de Vaux is the first to mention this chapter [Carra de Vaux (1900), p. 29] (cf. p. 63):

Codex Hagia Sophia 2755.

In total there are 9 apparatuses in this part and most of them are

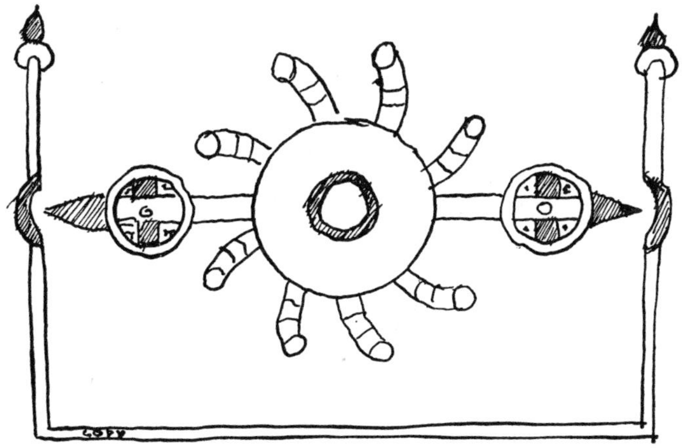

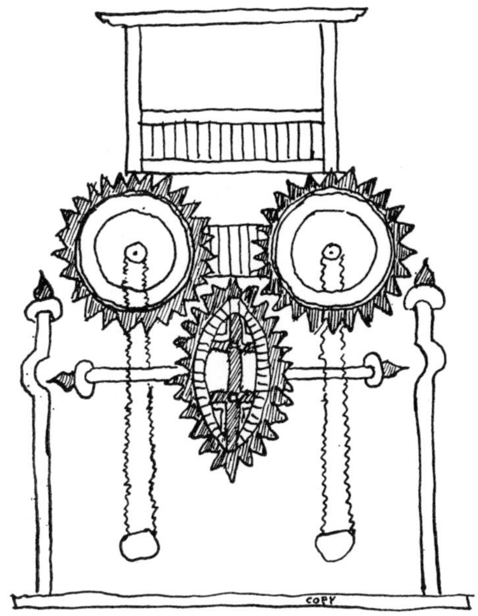

Fig. 51 b and Plate 2. Oxford 954. The first drawing shows a perpetual-motion wheel worked by mercury. The second is also a perpetual-motion wheel but it is also impossible to relate this to the text.

for perpetual motion. The last one is a contrivance to stop such wheels, *which turn by themselves.*

Three years later, Carra de Vaux (1903), p. 29, note 1 (cf. p. 63) mentions Codex Oxford 954 = Oxford March 669, which also has this chapter. Incidentally, this is the only chapter of the manuscript that has not been translated. He says of the apparatuses: "Most of them are obscure". See Figs. 51 a and 51 b.

Wiedemann [Beiträge *III*, Vol. 37, p. 231] writes as follows of Codex Gotha 1348:

"This chapter describes a whole series of illustrated devices." In a later paper [Beiträge *VI*, Vol. 38, p. 13], Wiedemann translates a few passages from this manuscript. Not until 1922 do we find a complete translation of the chapter, for which purpose the author has relied particularly on Leiden 499. Hans Schmeller: *Beiträge zur Geschichte der Technik in der Antike und bei den Arabern.* [Abhandlungen zur Geschichte der Naturwissenschaften und der Medizin, Vol. *VI* (Erlangen 1922), pp. 1–47].

Schmeller has reconstructed the machines on the basis of the text, but the drawings cannot have been of much use to him because text and illustrations are not in the same sequence. Even with the aid of *Istanbul* 1884, the sequence of which is slightly different, it is impossible to relate the pictures to the text, and it is quite impossible to decipher the constructions. See Fig. 52.

Schmeller notes that *Oxford* 954 deviates somewhat from *Leiden* 499

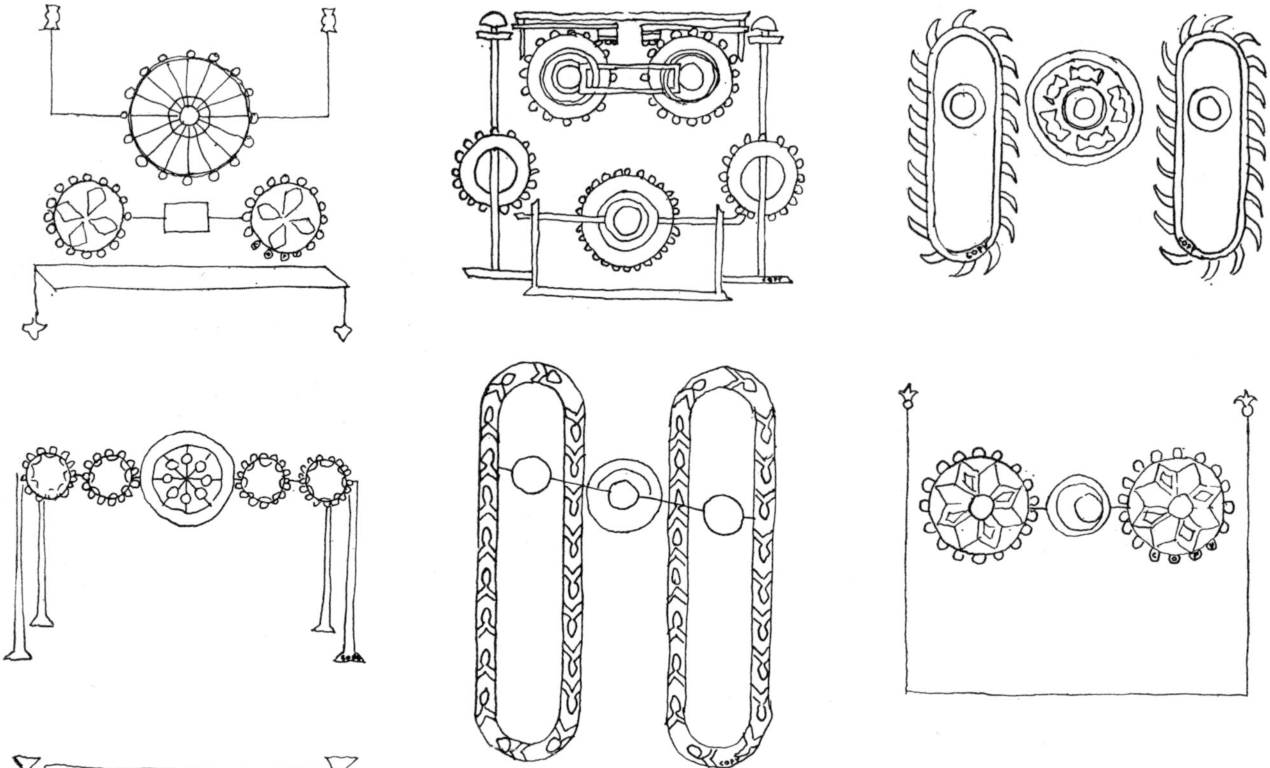

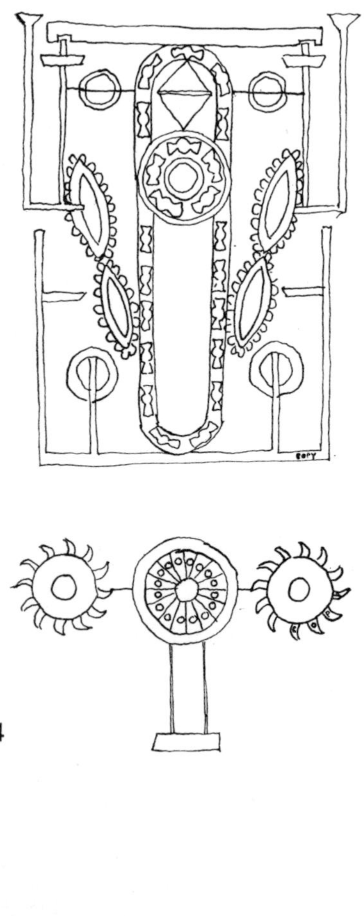

Fig. 52 and Plate 4. Istanbul 'Asʿad 1884. Drawings from "Wheels which turn by themselves". The drawings are purposely made unintelligible because they are perpetual-motion machines.

and *Gotha* 1348, and we encounter some of these deviations in *Istanbul* 2755. Mr. A. Zaqzūq, Keeper of the Museum of Ḥamā, has therefore given me a partial translation of this manuscript. See Fig. 53.

Although more systematically arranged, the text is by no means intelligible. There are eight machines and a supplement in two sections. One of these – and the only one capable of interpretation in the whole chapter – deals with ways of stopping *wheels which turn by themselves*. The other is about a complicated well sweep, which is also to be found in Schmeller.

Each Figure is introduced by an account of the application of the machine.

1. The First Figure: A wheel which turns by itself and lifts up water or turns a millstone.
2. The Second Figure: A wheel which turns by itself and lifts up water or turns two millstones or two wheels (dawlāb).

3. The Third Figure: A wheel which turns by itself and causes the turning of eight millstones or wheels. There are steps between the two edges (*i.e.* a lantern wheel).
4. The Fourth Figure: Making a wheel which irrigates by itself or turns a very big millstone. (The translation is given below).
5. The Fifth Figure: Making a wheel which turns by itself and turns four millstones or more.
6. The Sixth Figure: Making a wheel which turns by itself and thus turns a millstone or a wheel. (Also translated below).
7. The Seventh Figure: Another wheel which turns by itself.
8. The Eight Figure: Making a machine which turns by itself. This irrigates in a day and a night thirty . . . ?
9. How the wheels are stopped during their turning. (A translation is also given for this device).
10. A kind of wonderful circle.

Fig. 53 and Plate 1. Hagia Sophia 2755. Wheels which turn by themselves. The text describes various perpetual-motion machines coupled to potgarlands. The drawings are incomprehensible.

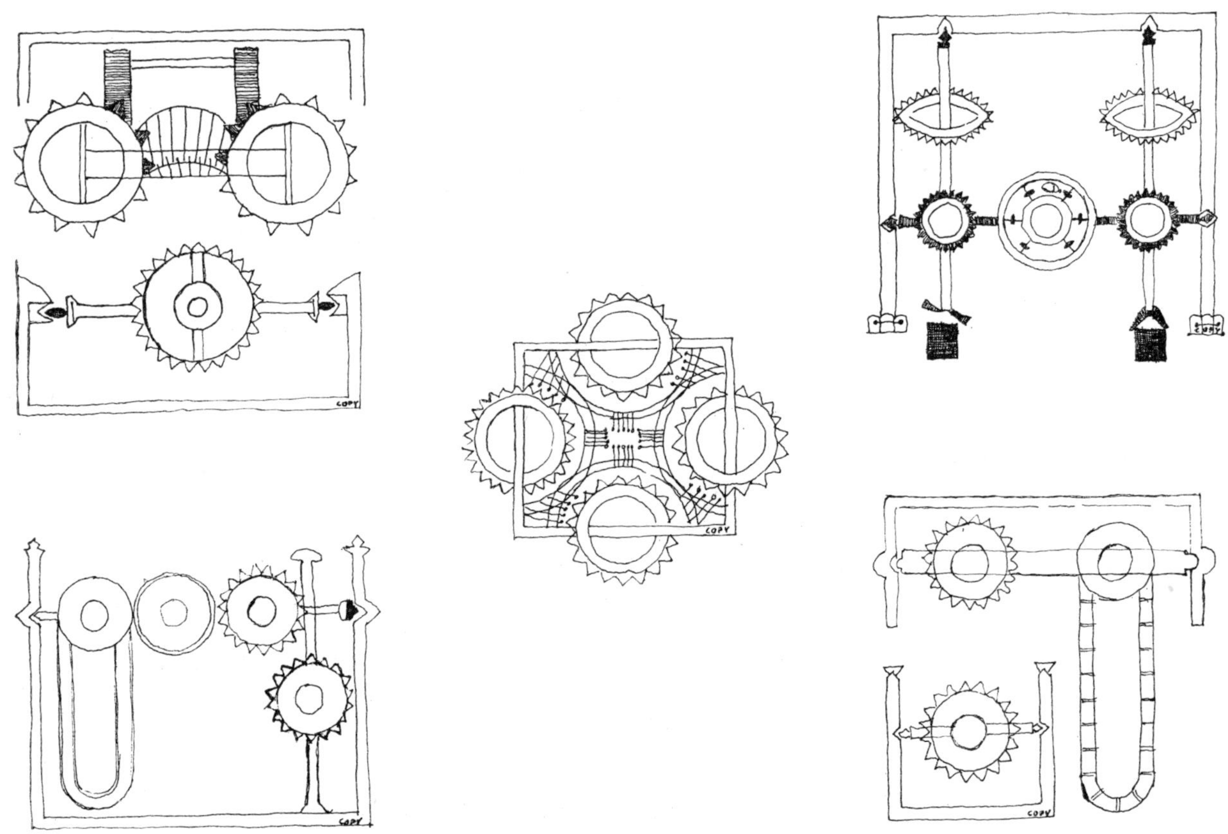

Islamic Waterwheels

The introduction is followed by a technical description of the machine, which is far less detailed in this manuscript than in that used by Schmeller. Some of the designs are undoubtedly the same in both manuscripts, but it seems as if all dimensions have been halved in one of them, which makes comparison difficult.

Two of the eight Figures – and only two – mention mercury. These are Figures Fourth and Sixth.

Arabic

The Sixth Figure.
Making a wheel which turns by itself and thus turns a millstone or a wheel (dawlāb). It is stated (made) in this regard (way):

1) A strong circle (wheel) of hard wood, its diameter being five (15?) *spans*. It is mounted on a strong pole (shaft).
2) Twenty four towers (tubes) – (*Oxford* 954 has 24 towers too) or boxes are mounted uniformly on it.
3) Mercury is poured into these boxes according to their sizes.
4) Boards of teak wood are mounted on the circle (wheel) on the two sides and along the rim. It (is) then painted with asphalt so that the circle (wheel) is 15 *spans* wide. It should be stationary (balanced?).
5) Two small circles (wheels) are mounted (fastened) to the sides of the big one. (We have three wheels sharing a shaft).
6) Buckets like those of a camel (a potgarland) are added to the small circles.
7) The weight of the mercury makes the circle turn, after releasing it, that is, after taking off the device which stops it.
8) It irrigates from 30 to 50 . . . daily.

Schmeller has more or less the same text for his machine No. 8. Both Schmeller and Lynn White Jr., p. 130 (cf. p. 55), note that this text corresponds in many respects to a text by the great Hindu astronomer and mathematician Bhāskarā, dec. 1150 [Bibliotheca Indica *XXXII*: *Hindu Astronomy*: Siddhānta śiromaṇi. tr. L. Wilkinson, Calcutta 1861, pp. 227–228].

Hindu

Instruments which turn by themselves

1) Make a wheel of light wood and along its circumference put
2) hollow rods all having bores of the same diameter, and let them be placed at an angle somewhat inclined from the perpendicular; then half fill these hollow rods with mercury: the wheel thus filled, when placed on an axis supported by two posts, will turn by itself.

Bhāskarā mentions yet another instrument, which is not included in the text used by Schmeller, but which Lynn White Jr., p. 131, has discovered in an anonymous Latin work of the late fourteenth century. However, he lacks the Arabic link.

Hindu

Here, first, is the Indian text, which reads as follows:

1)* Or scoop out a canal in the rim of a wheel; then plastering leaves of the *tála* tree over this canal with wax, fill one
4)* half of this canal with water and the other half with mercury till the water begins to come out, and then seal up the
5)* orifice left open for filling the wheel. The wheel will then revolve of itself, drawn round by the water.

The missing link is to be found in *Istanbul* 2755, although, here, as the source of power for a machine:

Arabic

The Fourth Figure.
Making a wheel which irrigates by itself, or turns a very big millstone.

1)* A hollow circle (wheel) of copper or iron, its diameter being five cubits. Its rim is covered with leaves, and its thickness is equal. It has a hole for the hand. It is heated on fire, and covered with asphalt to block all the holes.
2) Thick wooden boards are placed on it.
3) It is mounted on a pole or axle.
4)* It is filled with mercury to its quarter and then water is poured over the mercury to reach the hole, which is then tightly locked.
5)* It is turned strongly at first. Owing to the heaviness of the mercury and to the fact that it is placed in one of the parts, the wheel turns.
6) Some axles or teeth are mounted to this wheel which turn a wheel (dawlāb) or a millstone.

On the basis of his knowledge of the perpetual rotation of the firmament, the Indian astronomer designed a number of instruments with the same capability. An Arab then found it more expedient to let these instruments act as sources of power for mills and waterwheels.

In order to strengthen the reader's belief in the infallibility of these machines, the chapter concludes with an easily intelligible section on how to stop such perpetual-motion machines:

Arabic

How the circles (wheels) are stopped during their turning.
If you want to stop the wheels do the following:

1) Make an extra part about one *span* long to the side of the pole.
2) Half of this extra length should be square and (the outer) half should be circular, as it is shown in the illustration below.

(There is space for such an illustration). The circular part should be pointed.

3) Take a piece of iron one *cubit* long and one small *span* wide. Make a square hole in this and place in it a strong string between the pole (the piece) and the earth with two extra *cubits*.
4) Make for each wheel a string like this having a piece of iron in its top.
5) Ask a man to stand on the other side of the wheel (the other end of the axle) and then both of you place the end(s) of the axle in the square hole(s). Your position must be to the direction in which the wheel turns; not to the direction of turning.
6) Each of you should have a hammer to strike on the piece of iron, so that the square part of the pole will enter into the hole. Then the circle will stop even if it has ten thousand pounds.
7) If you want to turn the wheel once more, support it with a wooden crossbeam, then pull out the piece of iron and the string, then take off the supporting beam, then it will turn.

This last part gives a hint of the author's profession: he is either a crane builder or a turret clockmaker, since he knows that it is dangerous to stand on the wrong side of the crank handle of a hoisting winch.

It is, in fact, extremely difficult to comment on such a chapter as *Wheels which turn by themselves*, but before we judge the authors too harshly, we should remember to whom these manuscripts were addressed. They were intended for sultans, not for the simple carpenter; and this may be one of the reasons for the tendency towards sensationalism – the science fiction of those days – rather than towards realism. In any event, the next two contributions seem to indicate that these ancient designers could also produce extremely practical results when they had the peasant's waterwheel in mind.

Miniature from Baghdad

12th century
Baghdad

Location: Bibliothèque nationale, Paris.
Catalogue entry: Fonds Arabe 5847, folio 69.

The best photograph of this illustration is to be found in *Histoire général des techniques*; ed. P. Ducassé. Paris 1962, facing p. 384. The picture shows a special type of noria, and from it we can see the following: at the top we have a palm-trunk resting on two brick-built, stepped pillars. The upright shaft is provided with a gudgeon and an iron ring. We see that the oxen trot the "right" way round, but the artist has unfortunately located the scoop-wheel on the wrong side of the upright shaft – it should be between the oxen and the shaft. In the walled circular track there is a large outlet, from which water is led to ditches in the garden.

Fig. 54a. Spiral scoop wheel from the 12th century. Miniature from Baghdad [*Baġdād*]. *Bibliothèque nationale, Fonds Arabe No. 5847, folio 69 v. The system is still used in Egypt, particularly in the Nile Delta.*

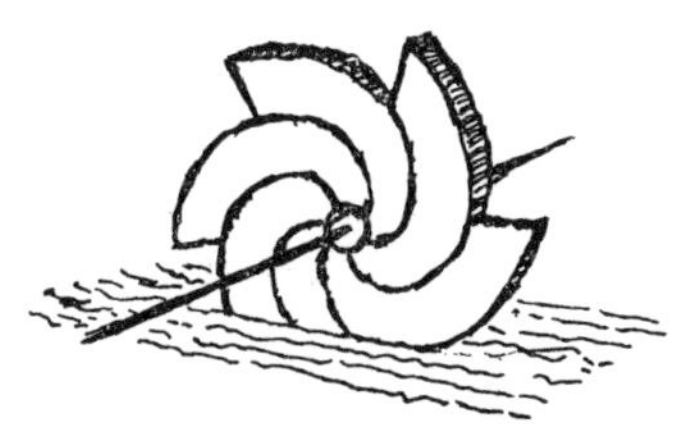

Fig. 54b. The spiral scoop wheel.

One of the most difficult problems in water-lifting engineering is that of raising large quantities of water a small lift. If it is found necessary to raise the water twice the required height, then the cost is also doubled. It is precisely this problem that the design shown manages to solve. The spiral scoops raise the water to ground level – not, as in the case of the Spanish noria – far above this. This scoop-wheel therefore operates with a very high degree of efficiency. Engineers at a research laboratory north of Cairo are still trying to improve the shape of the scoop in order to achieve optimum output. The machine is very popular in Egypt and seems very modern in its design. As will be seen in Figs. 54 a and 54 b, this is not actually the case, the design having been known as long ago as the 12th century. In other words, this problem was recognized – and solved – in earlier ages, so it was apparently not only perpetual motion that interested designers at that time.

Ibn al ʿAwwām

12th century
Spain

The most interesting record of the Spanish noria is to be found in the book of ibn al ʿAwwām, from the beginning of the 12th century, which is entitled *The book of agriculture.*

The references are:

Abu Zacaria Iahia: *Libro de Agricultura.* ed. Josef Antonio Banqueri. Vol. *I.* Madrid 1802, pp. 146–147:

> Dice Abu-el-Jair que en (cada) estado (ó cada dos varas) de la maroma de la noria haya cinco arcaduces (cubos ó cangilones), y que quantos más fueren los dientes en la rueda pequeña que juntamente con la mayor hace circular la dicha (maroma) vendrá a ser esta máquina más liviana y ligera, y lo mismo si el palo atravesaño fuese largo; el qual no perjudica sea de treinta ó casi treinta palmos. Que también se da fácil (curso á la maroma) doble con cortar del palo derecho, lo que sobresale por encima del horado del atravesaño; y lo mismo si las rodajas que llevan (ó donde estriban) los arcaduces fueren de madera pesada; á cuyo fin han de hacerse más gruesas y pesadas de lo acostumbrado, respecto á que así se logra aquel efecto. Dicen, que para estorbar el encuentro de las gradas de la escalerilla con los arcaduces en el agua del pozo se haga á cada uno de estos en el fondo un pequeño agujero, (con lo qual) no ladeándose, estan libres de quebrarse unos con otros, ó en la expansión del pozo; (los quales) vaciándose también al parar la maroma, por esta causa dura la misma mucho tiempo.

J. J. Clément-Mullet: *Le livre de l'agriculture d'Ibn al-Awam,* Vol. *I.* Paris 1864, p. 129:

> Abou'l-Khair dit qu'il doit y avoir, par chaque *qāmah* (brasse) de la corde de la noria, cinq godets environ. Il ajoute: Toutes les fois qu'on augmentera le nombre des dents de la petite roue qui fait tourner, c'est-à-dire qui donne le mouvement à la noria, et que l'on augmentera (le diamètre de) la grande, la machine fonctionnera avec plus de légèreté et plus d'aisance. L'axe ou arbre de couche par sa longueur, ajoute aussi à la facilité (de la marche) de la noria; il n'y a aucun inconvenient à ce qu'il soit de trente *šabr*es (6.93 m) ou environ. Un des moyens de rendre le mouvement encore plus facile, c'est que l'axe, dans la portion qui est au-dessus du trou dont il est percé, reçoive (litt. soit coupé par) une pièce de bois disposée perpendiculairement. Un autre moyen d'allégement, c'est que la roue (litt. le cercle) qui supporte les godets soit en bois dur très-pesant, et qu'on lui donne plus d'épaisseur et de poids qu'on ne le fait habituellement. On a encore recommandé de pratiquer un petit trou au fond des godets, comme moyen de les empêcher de se heurter dans l'eau contre la rampe; on prévient encore, par là, un mouvement de torsion dans l'eau et la fracture des uns par les autres, ou contre le

> mur d'enceinte du puits, et quand la noria s'arrête (et cesse de marcher) les godets se vident (de l'eau qu'ils contiennent), et par suite la corde durera plus longtemps (litt. la vie de la corde sera prolongée,) Dieu aidant.

The same translation appears in:

G. S. Colin: *La noria marocaine .....* [Hespéris, Vol. *XIV* (Paris 1932), pp. 22–60] and in Caro Baroja (1955), p. 29.
E. Wiedemann: *Beiträge zur Geschichte der Naturwissenschaften X.* [Sitzungsberichte der Physikalisch-medizinischen Sozietät (not Gesellschaft) in Erlangen, Vol. 38 (Erlangen 1906), p. 334].
Th. Schiøler: *Virkningsgraden af en maurisk noria på Ibiza.* [Ingeniør- og Bygningsvæsen, Vol. 56 (København 1961), pp. 263–264].

We now turn to the French version of the construction of the noria:

> ....abu al Ḫayr says that there should be about five pots for every *qāmah* (167 centimetres) of the noria-rope.

Caro Baroja (1955), p. 29, includes the most important Arabic terms in brackets. The text refers to the sāniya, which is here correctly translated noria. Banqueri gives a *qāmah* as equal to two *varas*, a *vara* being an old Spanish unit of lenth equal to 83.5 centimetres. Caro Baroja gives it as equal to the height of a man, and Clément-Mullet uses the term *brasse*, which is approximately 1.62 metres.

As the normal height of a pot is about 30 centimetres, this means that the pots must be lashed closely together. See G. Dalman (1932), fig. 48 (cf. p. 41).

The next sentence in the chapter by ibn al ʿAwwām is difficult:

> Every time the number of teeth is increased on the little wheel that turns the noria (the potgarland wheel) and (the diameter of) the big one is increased, the machine works with greater lightness and versatility.

The Spanish translation is almost the same, but no mention is made of increasing "the big one". The meaning of this sentence is not altogether clear, but it seems to relate to the meshing. It is correct that an increase in the number of teeth on the little wheel will reduce the variation in the speed ratio between the two wheels, but we have to know the teeth ratio and the diameter ratio before we can treat the problem theoretically.

> The traction pole or the sleeping tree, depending on its length, facilitates the working of the noria. A length of up to about 30 *šibr* [*šabr* sing. *šibr* pl.] is suitable.

The meaning of this must be: the force required to turn the noria depends on the proportion between the radius of the lantern wheel and the horizontal projection of the pole; the longer the pole, the lower the force that is required, and vice versa. We are told that the pole can have a length of up to 30 *šibr*. According to Clément-Mullet, this is equal to 6.93 metres, and the Spanish version gives practically the same figure.

Normally, even in antiquity, the diameter of the circular track was about 5 metres, but here it is the radius that is discussed. Clément-Mullet says of this sentence: 'The translation of this passage presents great difficulties.' Looking at the problem from another angle, if we examine a noria, it is easy to find the diameter of the circular mule-track. This in turn is the projection of the pole. The length of the pole has no theoretical value because the pole is often bent (Fig. 55). The meaning must therefore be: The circular track may have a diameter of up to 7 metres.

The next sentence is far from clear in the French translation and totally incomprehensible in the Spanish version:

> One of the ways of facilitating the movement is to insert a piece of wood perpendicularly in the axle (upright shaft) in the place above the hole for the traction pole.

The object in question must be the little guide-rod, offset at 90° in front of the traction pole. It is correct that it is placed above the traction pole. See Fig. 55.

The technical meaning of the next sentence is not quite clear:

> Another way of facilitating the movement is to make the wheel (the circle) [singular], which supports the pots, of hard and heavy wood and more thick and heavy than usual.

The Spanish version is different:

Fig. 55. Noria, Portugal.

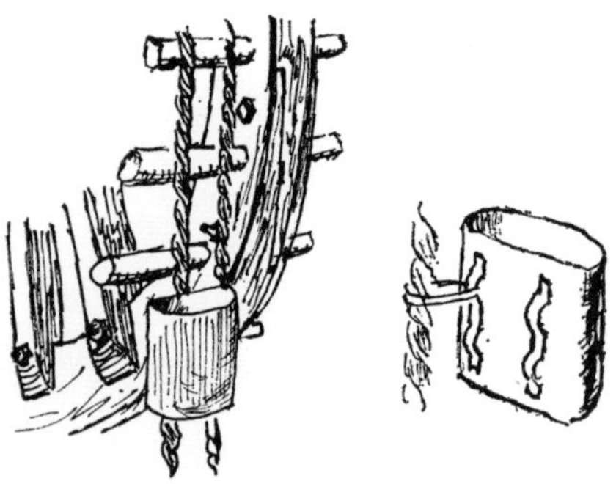

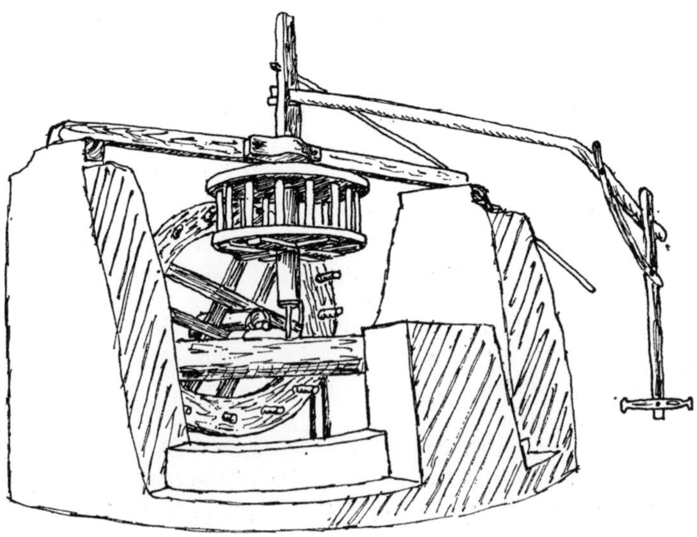

And the same [to facilitate the movement] if the rollers (las rodajas) [plural], which support the pots, are made of heavy wood ......

If this plural is correct, it must be the teeth that support the potgarland. The meaning is then: the teeth should be made of hardwood, unlike those in the lantern wheel, which are made of softwood. This interpretation is based purely on the plural form. A close examination of the Arabic text is necessary.

The last sentence is clear.

It is also recommended to make a little air vent in the bottom of the pots to prevent them from bumping against the ramp in the water, and from turning in the water, or from breaking by colliding with each other or the wall of the well.

As a ramp has a slope and as, furthermore, Wiedemann uses the term *Geländer*, the ramp must be the two guide-rods.

As the noria stops, the pots empty, and this increases the life of the rope.
If 'Allāh permits.

As will be seen from the foregoing, it is not possible to obtain a precise meaning from many of ibn al ʿAwwām's sentences, but the text is still interesting. It will be noted that the passage does not deal with the noria as a new construction, but specifies methods of improving a design that is already familiar. Some of the improvements aim at increasing the lifetime of the pots and ropes, while others aim directly at improving the efficiency of the machine. Of course, the craftsmen of that time did not fully understand this concept, but they were instinctively working in the right direction.

Vignette No. 4. Paternoster chain and treadwheel in ms. Marsh 669 folio 16 r.
Die Leute von Iṣfahān.

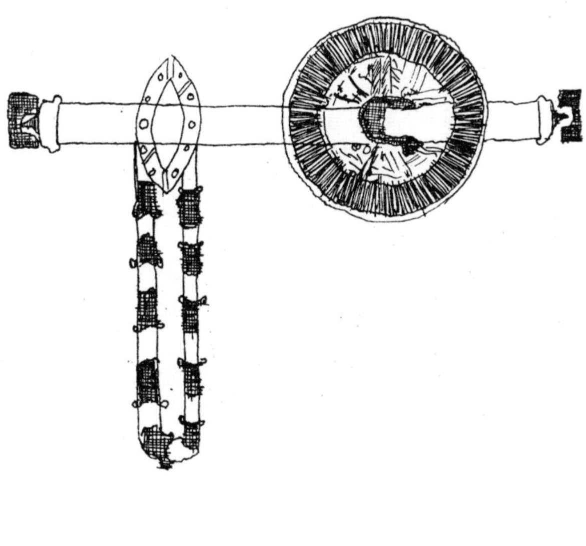

CHAPTER 4

Water-Raising Machines in Indian Texts

Literature from B.C. 300 to A.D. 1300 reviewed in inverse chronological order

This subject has been dealt with by Berthold Laufer: *The Noria or Persian Wheel.* (Oriental Studies. In honour of Cursetji Erachji Pavry, ed. J. D. C. Pavry, Oxford 1933, p. 238–250.)
The above article has been commented on by Ananda K. Coomaraswamy: *The Persian Wheel.* [Journal of the American Oriental Society, Vol. 51, 1932 pp. 283–284]:

Fig. 56. A waterwheel turned by the hands and feet. 18th century miniature. India.

Fig. 57. A waterwheel turned by the hands. 9th to 10th century relief. Maṇḍor, India.

> The well-known device consisting of a series of vessels bound sideways on a wheel, and so used to raise water from a shallow well as the wheel is made to revolve by means of *a geared shaft worked by oxen* (italicized by the author), usually for irrigation purposes, is a familiar sight in northern India, and representations in Mughal painting are not uncommon.

A sketch of the painting in question is given in Fig. 56, but it shows only a potgarland wheel without gear or oxen. Coomaraswamy also refers to a bas-relief. See Fig. 57. This, too, shows a potgarland wheel turned by the hands and feet.

With a view to the interpretation of the Indian texts, it is necessary to introduce yet another water-raising device that is really outside the scope of this book. This device consists of a large pulley suspended from a wooden framework over a well. See Figs. 58 a, 58 b and 59. A rope runs over the pulley with a big leather sack attached to one end and a draught animal to the other. A second rope makes it possible to empty the leather sack automatically when this reaches the top of the well. There is a good drawing of this pulley-machine in F. Christiansen-Weniger: *Alte Methoden der Wassergewinnung für Bewässerungszwecke im Nahen und Mittleren Osten* ... [Wasser und Nahrung, Vol. 6 (Düsseldorf 1961), p. 30].

Fig. 58 a. Detail from an Indian miniature from 1562. This water-raising apparatus consists of a large leather sack drawn up from the well over a pulley-wheel. The system is older than the sāqiya, but is still widely used in India, Arabia and Algeria. See also Figs. 58 b and 59.

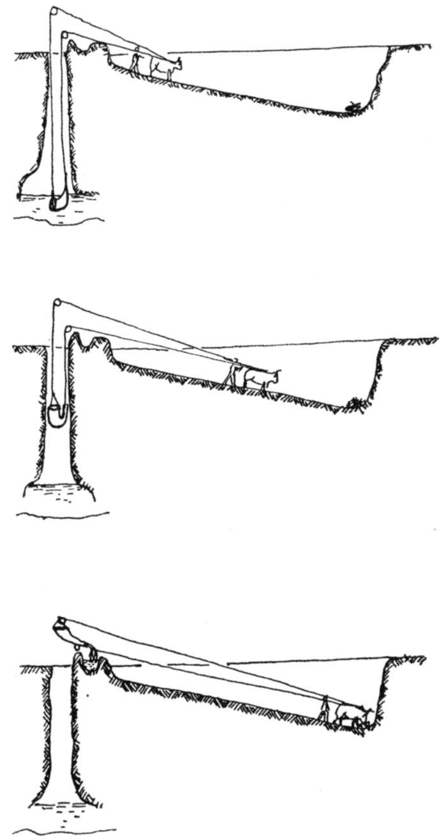

Fig. 58 b. Supplement to Fig. 58 a.

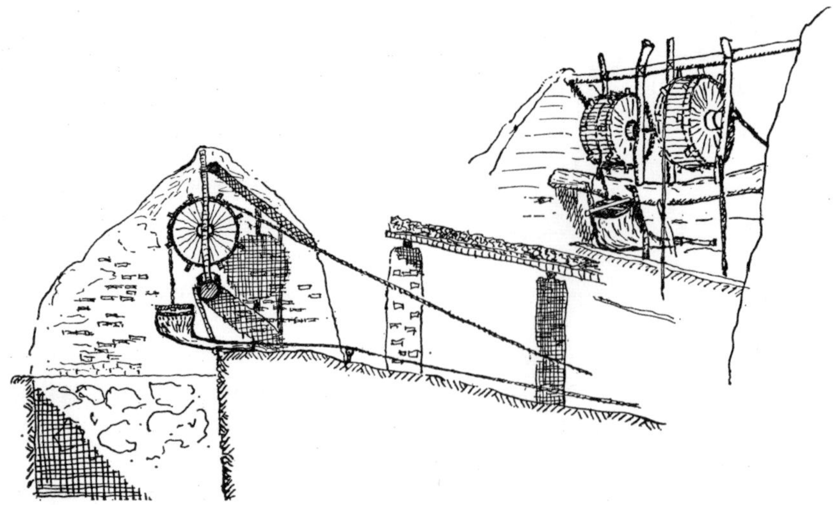

Fig. 59. Water-raising apparatus consisting of a large leather sack drawn up from the well by means of a main rope. A second rope makes the sack empty at the appropriate time. Persia.

All but the last of the following texts are taken from Coomaraswamy's article. I am indebted to Professor H. Hendriksen for valuable information on these texts, and for his careful interpretation. The work of interpreting these fragments is difficult, firstly, because the original text has been changed by later commentators, and secondly, because the definitions of the various machines given in dictionaries are not easy to understand. However, the information contained in the texts does, to a certain extent, permit us to decide whether the machine is a sāqiya or not.

1306 *India* Prabandhacintāmaṇi was written in the year 1306, and one of its sentences suggests a potgarland machine, *i.e.* a sāqiya, because it mentions field-irrigation and talks of buckets in the plural form.

> Do you not see that in the waterwheel for irrigating fields, the empty buckets become full, and the full buckets empty?

This seems to me to be the only text that really relates to a sāqiya.

11th century *India* In the next text, we encounter the term, araghaṭṭa, which is translated by V. S. Apte: *Sanskrit-English Dictionary*, (1924): "A wheel or machine for raising water from a well (Marathi rāhāṭ). It usually consists of a single wheel with spokes on each side serving as handles to turn it, and a rope with a bucket attached to it passes over this wheel." This choice of words indicates a windlass similar to that shown in Fig. 60.

Māhārāshṭrī, 18/19, which is to be found in H. Jacobi: *Ausgewählte Erzählungen in Māhārāshṭrī* .... Leipzig 1886, has a story about an āraghaṭṭika.

Fig. 60. Egyptian windlass, 1964.

.... a man who works a Persian wheel.

It is impossible to reach a decision about this machine.

11th century
India

Coomaraswamy, p. 284, writes:

> In a verse cited by Kṣemendra from Amaraka, quoted in the Journal of the Royal Asiatic Society 1916, p. 168, there is a simile of the water that rushes out of a bucket as it turns.

It is difficult to decide what type of machine this simile reflects; it may simply refer to the individual pot in a potgarland.

7th century
India

In the following references, the term udghāta occurs, which means lifting or raising, while ghaṭīyantra means bowl-machine.

Coomaraswamy mentions three passages from Harṣacarita, which are to be found in Nirnaya Sāgara, ed. 1925. Professor Hendriksen has given me an interpretation of the first passage, which differs somewhat from that proposed by Coomaraswamy:

> The earth has twisted hair (as worn by holy men in India) because of the braids of cummin that are watered to all sides by the bowl (or bowls) of the well-wheel.

The next reference, p. 104, is less ambiguous:

> With his right hand he turned the rosary garland just like the garland of a well-wheel that raises all sweet taste from the well.

There seems to me to be no doubt that the machine referred to here is similar to that shown in Fig. 57.

The last passage from this work refers to a machine that works both day and night, p. 254:

> The birth, and the ageing, and the death of man is like unto the rope of the well-wheel, which moves round both by day and by night and which grows longer and longer.

This must refer to a machine powered by oxen, but even though a pot-garland may stretch with time, we cannot be sure that the passage refers to a sāqiya.

B.C. 350 or A.D. 400
India

There is one text from this period dealing with various water-raising devices and their accessories. The text is mainly concerned with the disadvantages of certain methods and possibilities for improving these. It is not my impression that the alterations are in any way radical, even though some of the comments mention potgarland machines and even machines that are worked by elephants. The text in question is to be found in T. B. Horner: *The Book of the Discipline*, Vol. *V*, London 1952, p. 168 (Vinaya *II*, 112 (Cullavagga chapter, *V*, 16.2)):

> Now at that time monks drew water by means of jungle rope and by means of a waistband.

As will be seen from the following, this was found to be a poor solution:

> I [Buddha] allow you, monks, a cord for drawing water. Their hands became painful.

However, this improvement does not seem to have been satisfactory either, for we then read that

> I allow, monks, 1) a well-sweep, 2) a hand-wheel, 3) a wheel and buckets.

And now it appears that the pots have been broken:

> Many vessels were broken. I allow, monks, three (kinds of) jars: a copper jar, a wooden jar, 4) strips of animal hide.

Comments:

1) tulā = water-lifting device used by gardeners. (Beam of a balance.)
2) a hand-wheel karakaṭaka = "either having harnessed it to bullocks or having taken it with the hands, it is an instrument with a long strap for pulling up." This comment is from about the

year A.D. 500, although it is based on a considerably older comment that has since been lost.

3) cakkavaṭṭaka = wheel and buckets. A footnote gives the translation as windlass. According to Professor Hendriksen, the word vaṭṭaka in cakkavaṭṭaka can hardly mean bowl or bucket. The translation "wheel and buckets" is based on another passage in Vinaya, in which there is a comment from the period about the year 500. Here, the original term – which can hardly include the meaning "bowl or bucket" – is replaced by a new term, arahaṭ(ṭ) aghaṭiyantaṃ = mechanical device with waterwheel bowls (pots?), *i.e.* as shown in Fig. 57.
4) A comment from about the year 500 reads: "a hide vessel that can be joined to the well-sweep or the hand-wheel."

Taking everything into consideration, it is doubtful whether the original text refers to anything but the simplest of water-raising devices that were known at that time and earlier, namely a well-sweep as shown in Fig. 87, a well with pulley, or a well with windlass, as shown in Fig. 60.

B.C. 450
India

However, as is apparent from another text from the same period, the people of that time were familiar with the water-raising device in the form shown in Fig. 58. The text in question is discussed by V. S. Agrawala: *India as known to Pāṇini.* A Study of the Cultural Material in the Ashṭādhyāyī, 2nd ed. The Tara Printing Works, India 1963:

> What is unique about this work is the fact that .... the Ashṭādhyāyī is the work of an individual author and not an accumulated result of additions made to it from time to time by generations of teachers ......
>
> p. 199 Agricultural implements:
>
> In Pāṇini *IV*, 2, 45, we have yuga and varatrā, which refer to the yoke and the thick long rope used for lifting water from a well, still called barat in western and bartā in eastern Hindi.
>
> p. 204 Irrigation:
>
> Irrigation with water from the wells was also known, as indicated by the word udañchana Pāṇini *III*, 3, 123, the large leather bucket used for lifting water, and also yuga-varatrā, "the yoke and the rope by which the bullocks were driven for raising the water". Pāṇini's own comment to *IV*, 2, 45.

None of the older texts indicates the use of a right-angle gear; in fact, not until 1306 do we find a text that suggests a more complex machine, possibly a sāqiya.

CHAPTER 5

Waterwheels in Ruined Islamic Castles

Brick building from A.D. 700 to A.D. 1300 reviewed in inverse chronological order

In the Islamic world we find a number of ruins which show evidence of a sāqiya having been incorporated in the building. In the following we shall take a closer look at some of these.

1313 *Cairo*

Between the Citadel in Cairo and the Nile there runs a great aqueduct, which appears to slope down towards the river. It is, in fact, a stepped aqueduct, the water from the Nile being raised up to the Citadel by stages at suitable intervals.

As mentioned, the water supply line starts at the Nile, where there is a hexagonal tower (al Ġūri's intake tower), which once housed six sāqiyas. Unfortunately, when I was in Egypt in 1963, I did not have the opportunity of inspecting the inside of the tower because the porter had mislaid the key. From this tower the water ran along its first stage to the next sāqiya, which raised the water to the next stage, and so on. See K. A. C. Creswell: *Early Muslim Architecture*, Part *II* (750–939), Oxford 1940, p. 258.

Approx. 1200 *Aleppo*

The Citadel at Aleppo had a particularly difficult water supply problem because water was needed for both the baths and the moat. The water for the baths was brought up by a sāqiya, while that for the moat came from a spring. Within the walls of the Citadel there are three very deep wells, one of which is covered by a vaulted roof. Beside this well there is room for a machine and a water tank. A Byzantine sarcophargus from the 6th century served as the water trough below the taps of the tank. However, the system itself is much later than this, presumably dating back to the 12th or 13th century. It is true that Ibn Ǧubayr (1185) speaks of the wells of the Citadel, but he makes no mention of the machines. Guy Le Strange: *Palestine under the Moslems*, London 1890, p. 364.

We find the plant on the right, directly after passing through the main entrance. The machine room itself measures 5.7 × 7.5 metres and is provided in the middle with a bedding stone for the upright shaft of the sāqiya. The diameter of the circular track cannot have been much less than 5.4 metres. In the vaulted roof there is a wooden beam with a bearing for the top pivot of the sāqiya. The upright shaft must have had a height of 3.6 metres, and I should imagine that the bucket-chain wheel itself had a diameter of less than 2 metres, because the well shaft measures 2.3 × 1.8 metres. It is difficult to decide where the horizontal shaft was placed, but all the same, there is no doubt that this was once a sāqiya with an elevated shaft, similar in design to the one shown in Fig. 27. The sketch in Fig. 61 gives an impression of the place as it appears today (1969).

Fig. 61. Sāqiya room in the Citadel at Aleppo.

The second well is on the left of the path, a little way past the main entrance. Here, there is nothing to see but the well shaft. Close by the long barracks dating from 1834 there is a two-stage well from the 13th century. This is 41 metres deep and measures 2.8 × 0.9 metres at the top, which seems to indicate that it, too, was provided with two sāqiyas, as in the case of Joseph's Well in Cairo. The well and its galleries have been described by R. Engelbach: *The Subterranean Passages of Aleppo Citadel.* [Ancient Egypt, (1920), pp. 107–110].

1176–1190
Cairo

A very well-known structure is Joseph's Well in Cairo. This is situated in one of the bastion towers of the Citadel. At the top, under a wooden roof structure, the beams of which are decorated with arabesques, there are (1963) the remains of a sāqiya. Halfway down the well there is yet another sāqiya, of which I have made a rough sketch. The big wheel was provided with about 50 cogs and had a diameter of 3 metres. The upright shaft was 2 metres high. The well, which is square, has been described several times, *inter alios* by F. L. Norden in 1755 (cf. p. 49). There is a comprehensive bibliography together with a number of references to older Arabic literature in Creswell, Part *II* (1940), p. 5.

876
Cairo

There is another building of interest, which Creswell calls The Aqueduct of Basātīn, at Fusṭāṭ, on the outskirts of Cairo. See: Creswell, Part *II*, (1940), pp. 329–332, plate 94, fig. 244:

> There must have been two sāqiyas, each with a vertical wheel over the shaft for the chain of buckets, and a horizontal wheel, with wooden cogs, geared into it. The latter would be turned by oxen walking round and round beneath it.

This description indicates a sāqiya of the elevated shaft type, but although Professor Creswell described the location of the aqueduct to me in person, I did not succeed in finding it.

Jordan

In the deserts of Syria and Jordan there are hundreds of ruins, all of which have had water supply problems that had to be solved. Most of the ruins have large tanks for rainwater, and almost all of them have a narrow well. All that has been necessary in the way of accessories has been a rope and a leather sack. However, in a few places, the Arabs have

Fig. 62. The hydraulic works at Quṣayr al ʿAmra, about 60 kilometres east of ʿAmmān in Jordan.

had sāqiyas, for instance at Quṣayr al ʿAmra, which is situated about 60 kilometres east of ʿAmmān. I visited this hunting pavilion, which lies deep in the desert, on Easter Sunday 1969.

705–711
Quṣayr al ʿAmra

The building, which is exceptionally well preserved, contains a bath – which indicates that large quantities of water were needed. The bath is in the Roman style, with oven and smoke ducts in the floor, but against the north wall, in accordance with Arab custom, there is a marble slab, the surface of which must be constantly covered with running water. This means that the water must be under pressure, as the top edge of the slab lies about 1.70 metres above ground level. For this purpose there was once an extensive hydraulic works here, about 6 metres from the bath, consisting of an elevated water tank, the bottom of which was located the same 1.70 metres above ground level. To supply the water tank there was a well and a sāqiya, the remains of which can still be seen. See Fig. 62.

The literature on Quṣayr al ʿAmra is extensive, but the principal work on the subject is to be found in A. J. Jaussen et R. Savignac: *Mission archéologie en Arabie.* (Publications de la Société des fouilles archéologiques, Paris 1909–1922). This work is in six volumes: *I Texte et atlas, II Texte, II Atlas, II Supplément, III Texte, III Atlas.* Volume *III Texte*, which is entitled *Les châteaux arabes de Qeṣeir ʿAmra, Ḫarāneh et Ṭūba*, contains a number of descriptions and ground plans, pp. 47–49 and 84–86. The volume entitled *III Atlas* has a number of good pictures of the hydraulic works: Plates *XVI*, 2; *XXXV*; *XXXVI*, 1 and 2; *XXXVII*, 4 and 5 and *XXXIX*, 1. Another publication that is worthy of mention is: Alois Musil: *Ḳuṣejr ʿAmra.* (Kaiserliche Akademie der Wissenschaften, Wien

1907, *I Textband*, fig. 44, p. 57 and fig. 45, p. 59, and *II Tafelband*, fig. *III*, *V* and *VI*).

There are ruins of sāqiya works at other desert sites as well; including Quṣayr al ʿAmra, there are seven in all. One of these is situated approximately 300 metres south-west of the pavilion at Quṣayr al ʿAmra (Musil, *Tafelband* figs. *I*, *II*), and I much regret that I omitted to go there, even though it is clear from my notes that I knew of the existence of two hydraulic works. I also knew that there were three other works at Quṣayr al Ṭūba, which is located 60 kilometres south of Quṣayr al ʿAmra, but these I was unable to visit for other reasons. Musil, *Textband*, fig. 98, p. 110, figs. 99 and 100, p. 111, fig. 101, p. 112, The drawing (fig. 101) is completely incomprehensible, as is also the text to this figure on p. 111.

The five works are shown in Fig. 63. At first glance, they appear to be different but, in fact, many of the dimensions are identical. Thus, the well is in each case framed by a square foundation, which is largely the same in all five instances. The distance from the centre of the circular track to the centre of the well is 5.5 ± 0.3 metres. In general, the buildings show a certain degree of standardization, but this need not be taken to indicate that the same architect designed them, since the principal form of the Palestinean sāqiya-works has remained unaltered through the ages. For example, the dimensions of the square columns at the works at Quṣayr al ʿAmra are the same as those shown in a photograph of a modern sāqiya in Georg Landauer: *Palästina, 300 Bilder*. München 1925, p. 134.

Fig. 63. Ruins of Islamic sāqiya works in Jordan. The top three are from Quṣayr al Ṭūba, and the bottom two are from Quṣayr al ʿAmra.

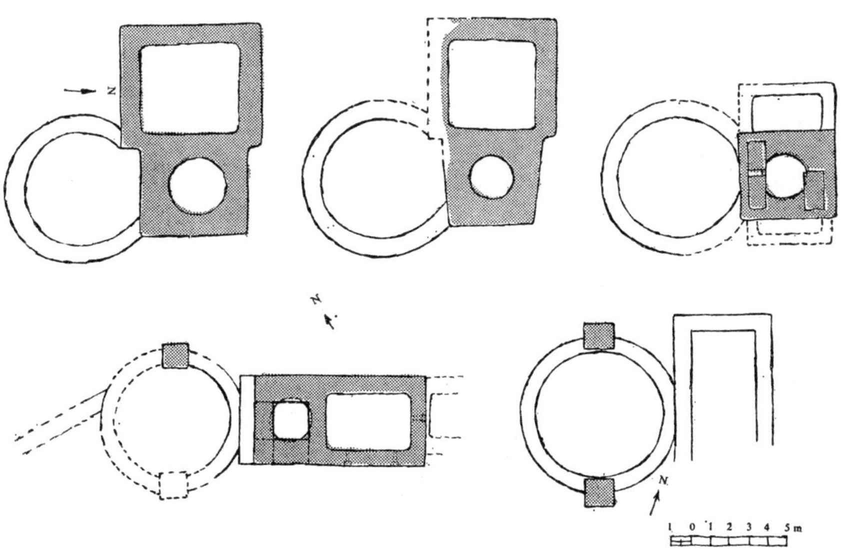

There is a sixth sāqiya-works in the Negev desert, which Musil, *Textband* p. 67, calls Ḳubbet el-Bīr, (31°02′N, 34°34′E). At this works there is a bath with a sāqiya. The well and the circular track have been almost completely destroyed, but the water tanks, which are on the roof, are intact.

The seventh works mentioned is located in the southern part of the Negev desert. Musil, *Textband* p. 67, calls the site el-Bīr near ʿAbde, (30°46′N, 34°45′E).

In the following we shall study a reconstruction of the sāqiya at Quṣayr al ʿAmra.

Report

The well is partly silted up, but still has a considerable depth. It is lined with sandstone ashlars, the curvature of which corresponds exactly to the diameter of the well shaft, which is 2.03 metres. However, the square housing above the well is smaller, the well-hole measuring only 1.9 × 1.7 metres. In other words, the paternoster wheel can hardly have been more than 1.3 metres in diameter and may not have been of the interior trough type, which requires a diameter of at least 2 metres. It is therefore reasonable to assume that the design was similar to that shown in Fig. 27.

Strangely enough, there was not a single fragment of pottery to be

Fig. 64. Reconstruction of the works at Quṣayr al ʿAmra.

found at the site. At all the other hydraulic works I have visited, there have been large quantities of fragments, so we must assume that buckets or leather sacks were used instead of pots at this particular well.

From the bucket chain we pass to the cog-wheel system, which is a somewhat simpler affair. As mentioned earlier, this is an example of an elevated tank construction, from which it follows that the gear was of the type that operates with an elevated shaft. Fortunately, one of the two columns is still standing; there are two recesses in this, which served as supports for the two horizontal beams. From these recesses and the remains of the circular camel track, I conclude that the sāqiya at Quṣayr al ʿAmra must have been broadly as shown in Fig. 64.

Today (1969), the two recesses mentioned above have been bricked up, and a wall has been built around the well to prevent animals and people from coming too close. There are a couple of holes in the old brickwork, which might indicate that there was once a wooden railing around the mouth of the well.

The water tank has three outlets, two of which contain fragments of

Fig. 65. Plan of the hydraulic works at Quṣayr al ʿAmra, surveyed by the author, 1969.

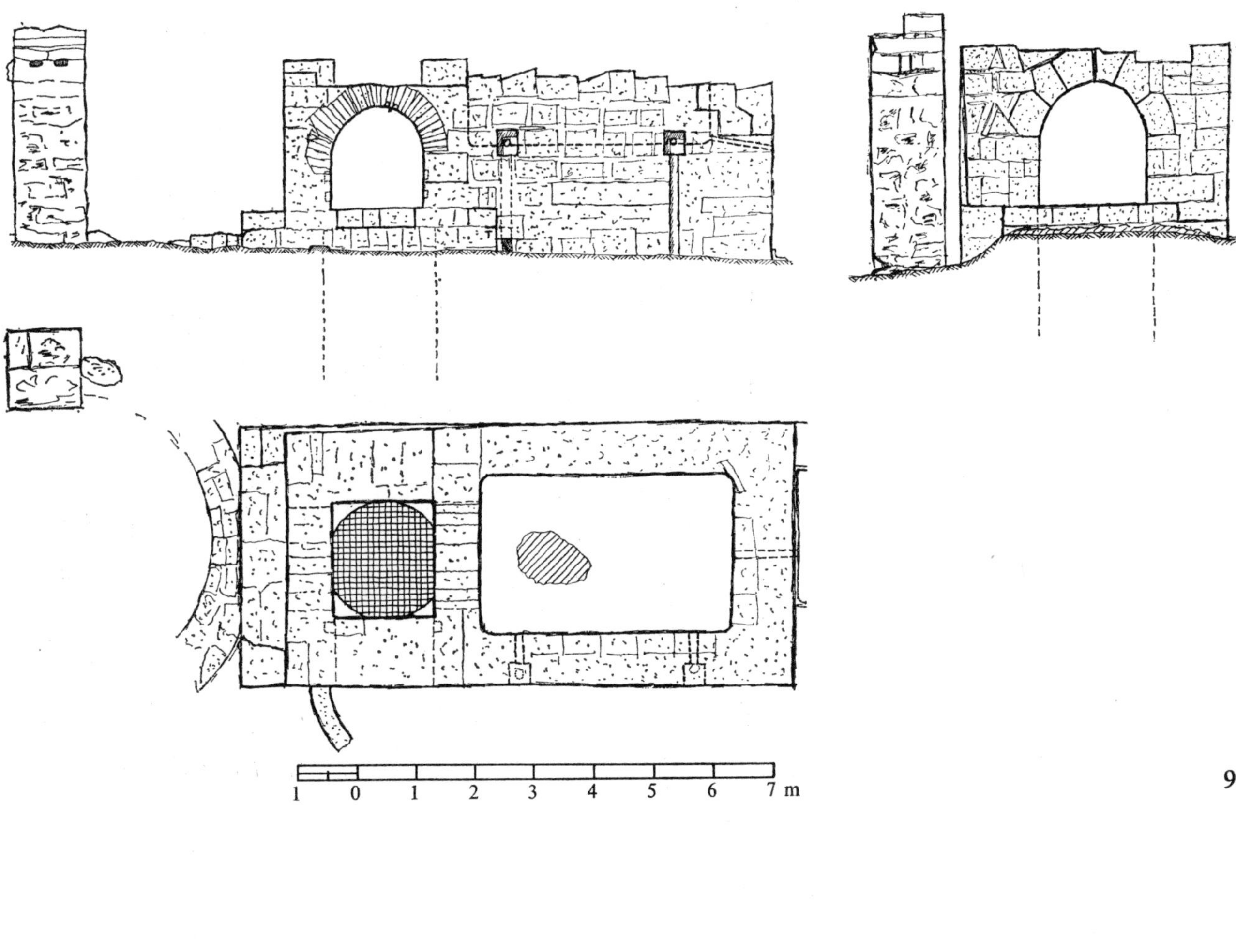

clay pipes. These pipes lead underground to the baths. The third outlet, which is on the eastern side, apparently leads nowhere, but when I brushed some sand away, I uncovered a layer of plaster, which proved to be the remains of a water tank. See Fig. 65.

There was nothing of particular interest in the first water tank, apart from the fact that it had a plaster coating. Neither in the well nor in the walls was there the slightest evidence of wear. There are usually heavy deposits of crustaceous lime around water tanks and wells, but I could find no trace of this either.

These few examples may easily give the impression that the Arabs used the sāqiya only in citadels and baths, but that is certainly not the case. The sāqiya situated 300 metres from Quṣayr al ʿAmra was used for irrigating a number of fields.

The use of the sāqiya in connexion with baths dates from Roman times down to our own. There is a drawing of an Egyptian public bath and sāqiya in Edmond Pauty: *Les hammams du Caire.* [Mémoires publiés par les membres de l'Institut français d'archéologie orientale du Caire, Vol. *LXIV* (Le Caire 1933), plate *VIII*].

Vignette No. 5. Thirteen century hydraulic noria. Codex Arab 368 folio 19 recto. Biblioteca Apostolica, Vatican. See Richard Ettinghausen: Treasures of Asia, Arab Painting. *Albert Skira 1962, p. 127.*

CHAPTER 6

Roman and Islamic Waterwheel Pots

Finds from A.D. 100 to the present day, reviewed in inverse chronological order

1. *Pots for the paternoster chain*

The pots used for the paternoster chain provide particularly interesting material for a study of the history of the waterwheel. Even today (1965) there are sāqiyas in use in Egypt, the earthenware pots of which have the same shape as those used in the Roman sāqiyas. Unfortunately, the material is very limited from a geographical point of view, since most of it comes from Roman Nubia, while only a small proportion throws light on the rest of the Roman world.

There are photographs of numerous examples from the 20th century. The drawing in Fig. 67 shows an example from the Valley of the Kings in Egypt. The pot in question is fastened to two sticks by means of two lashings, one round the neck of the pot and one round a characteristic knob on the bottom. However, there is another way of solving the problem of fastening.

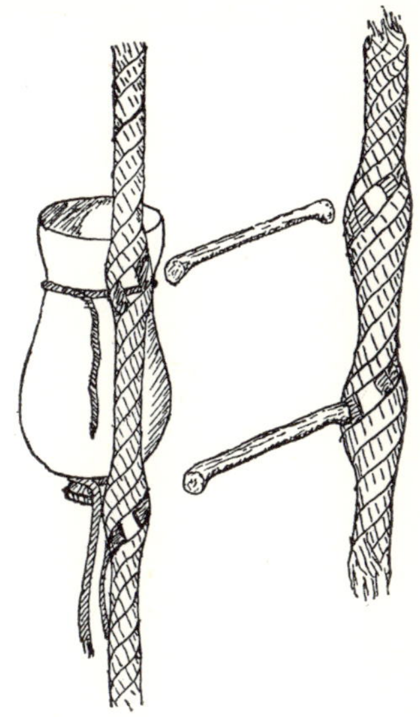

Fig. 66. The qādūs-pot, and the two small rods to which the pot is lashed.

Fig. 67

Fig. 67. Sāqiya near the Valley of the Kings, Egypt 1962.

7

Roman Waterwheel Pots

The drawing in Fig. 36, from the Province of Sind, shows how the pots can be fastened by means of a single lashing, but in this case, the pot must not be too tall.

If the pot is tall, the lashing must be placed close to its centre of gravity, as shown in Figs. 6 and 7.

Fig. 68

In the work mentioned earlier, by Ludwig Salvator: *Die Balearen* ... Leipzig 1897, Vol. 2, p. 347, there is a copperplate engraving of various unglazed ceramic vessels, see Fig. 68, including three different types of pots, each with its own characteristic shape. No. 1, which is best known in Majorca and Ibiza, can be traced back to the 12th century. No. 2, which is used in Morocco, is from approximately the same period; it can be seen, among other items, in Fig. 69 and glimpsed in Fig. 113. No. 3 is the classic waterwheel pot, the history of which can be traced far back in time.

Fig. 69

No figure

There is one example of the last-mentioned type in the National Museum in Copenhagen, No. G 1856, which was brought to Denmark by Captain H. M. Davidsen. Incidentally, this museum also exhibits a model of a sāqiya, made by an Egyptian sāqiya carpenter, and possesses a very big model of a Ḥamā-wheel, which was made in Ḥamā. There is a third model of an Egyptian sāqiya in the Science Museum, South Kensington, London, and a photograph of this is to be found in F. W. Robins: *The Story of Water Supply*. Oxford 1946, plate *III*, facing p. 44.

Fig. 68. Modern noria pots from Majorca.

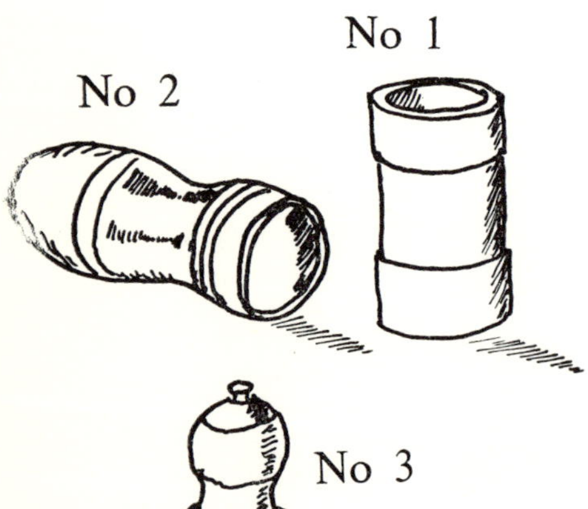

Fig. 69. Noria from Rabaṭ. The Kasba in Morocco about 1935. The construction is still kept in good repair. 1961.

Fig. 70. Modern sāqiya-pots from Egypt.

Fig. 70

État moderne (cf. p. 45), Vol. *II*, plate EE, shows a copper engraving of some unglazed ceramic ware, including three pots from the Egyptian sāqiya. See Fig. 70. In the same work, Vol. *I*, plate 73_4, there is a small scale drawing of a pot from the sāqiya at Joseph's Well in Cairo, which is reminiscent of the above-mentioned pot from Pakistan. (Cf. p. 44).

Fig. 71

The wheels at Hīt on the Euphrates (Fig. 32) are designed for slender pots, and I have found a similar pot, No. 29131, in the storerooms of the archaeological museum in Baghdad. See Fig. 71. The pot had been found near The Girls' School of this city. However, it was impossible to date it; it may have been only a couple of hundred years old. A total of seven pitchers were found there.

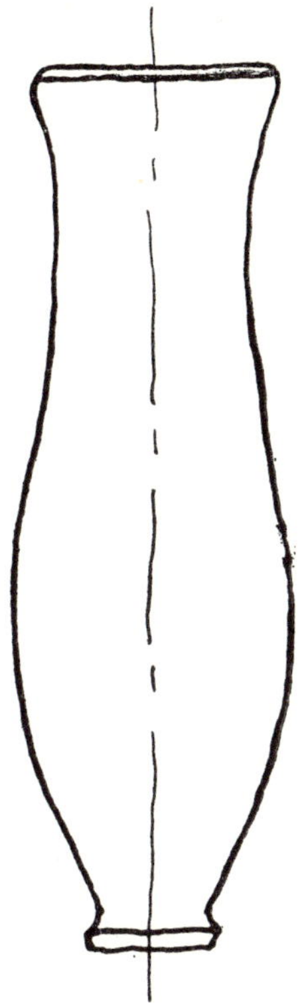

Fig. 71. Slender pot for hydraulic noria from Hīt on the Euphrates. Scale 1:5.

A similar slender pot is mentioned by T. E. Treptow: *Der älteste Bergbau und seine Hilfsmittel.* [Beiträge zur Geschichte der Technik und Industrie, Vol. 8 (1918), p. 179, fig. 45]. The pot was found in a mine in the Province of Córdoba and undoubtedly dates back to Roman times. See Fig. 72.

Fig. 72. Slender pot for a hydraulic noria from a Roman copper mine in Spain.

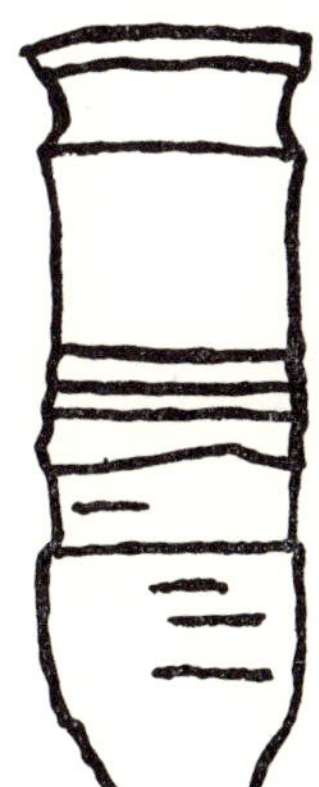

Fig. 73

In Fugmann's posthumous papers I discovered a note made by Prof. P. J. Riis during a visit to the Alcazaba Museum in Málaga. Enclosed with it was a little sketch of a Spanish noria pot. I visited this museum in 1971 and found not one, but four pots. See Fig. 73. I consider it very unlikely that the big one has been used for a noria as it is much too large for this pupose. Pot No. 183 was found inside the fortress walls at Torre del Homenaje. 12th to 14th century.

Fig. 74

This shape, with a narrowing waist, is known from a single find made during the Carlsberg Foundation excavation at Ḥamā in 1936. The pot, together with a full-scale drawing, can be seen in the National Museum in Copenhagen, No. [193]6 C 379. The pot is yellow and unglazed and, as may be seen in Fig. 74, it has no vent in the bottom. In this case there was no need of an air vent because the upthrust from the buoyancy of the pot exactly equalled its weight.

Fig. 74. Noria pots found during excavations at Ḥamā. From before 14th century. Scale 1:7.5.

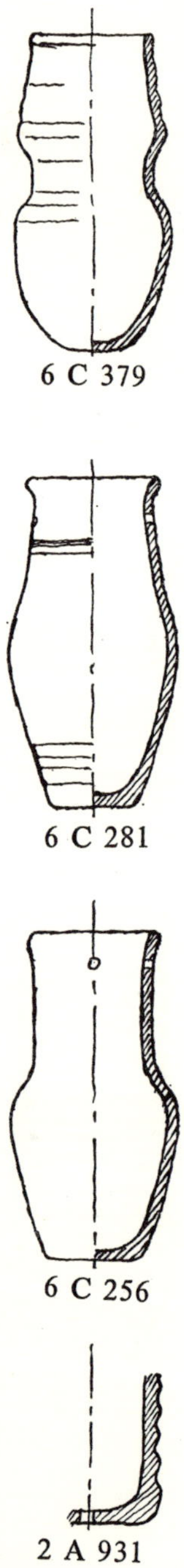

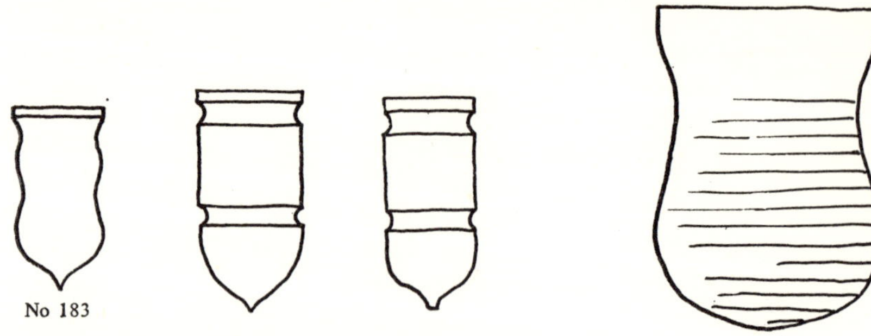

Fig. 73. Noria pots in the museum at Málaga. From the 12th century. Scale 1:10.

Another type of pot, which must also have been used for a sāqiya, was found during the same excavation. This type is particularly interesting because it is provided with holes just under the edge for the purpose of fastening. There are drawings of six of these pots in the National Museum, Nos. [193]6 C 290, 6 C 240, 6 C 281, 6 C 247, 6 C 256 and 2 D 944. The first four are almost identical. The last pot has two holes, and the others, four. See also: Fondation Carlsberg, Hama, Fouilles et recherches 1931–1938, *IV_2 – Les verreries et porteries médiévales*, par P. J. Riis et Vagn Poulsen et E. Hammershaimb, Copenhague 1957, fig. 998 (2D 944) p. 269.

Yet another fragment of the bottom of a pot from this excavation may be the remains of a waterwheel pot as it has a 10 milimetre vent in it. The inventory No. of this fragment is [193]2 A 931.

It is very difficult to date the pots from Ḥamā as they were all found in a large water tank that contained finds from all periods up to and including the 14th century.

Fig. 75 Unfortunately, waterwheel pots are rare in Syria, only two examples having been found, in the region of Aleppo. One of these is in the museum at Aleppo, No. 1064. See Fig. 75. The other, which was found "by a citizen in 1932", has been moved to the storerooms of the National Museum at Damascus, No. 4729. On the inside, it has a faint, greenish, algoid tinge, but there are no signs of wear. I have a colour photograph of this last pot.

Fig. 76 The Hayim Aveshalom Institute for Homeland Studies in Tel-Aviv has a pot that was found at Kefar-Manda in Galilee. In a letter to the author, Dr. Avitsur wrote as follows:

> .... Two years ago (1962) many such pots were discovered at the bottom of a silted well. They were relatively small. The pot holds 1.5 litres and the height is 21 cm. The biggest circumference is 40 cm. The pot has no marks. I think that the pot is from the 4th to the 6th

centuries A.D. Seen from beneath the knob has a sort of ornamentation. [The ornament looks like the spokes in a wheel.]

2. *Roman waterwheel pots from Egypt and the Sudan*

Unless a pot has the characteristic knob at the bottom, it is not always easy to decide whether it has been used in a waterwheel. From the cylindrical pots with broad grooves (Fig. 73), it is tempting to conclude that all pots with grooves come from waterwheels, but such generalizations are haphazard and, in this case, incorrect. The best way of ensuring that pots do indeed come from waterwheels is to find large numbers of them at the bottom of a well, and even then, they should preferably show signs of heavy wear and traces of algoid green.

Fig. 77

The first example we shall study is a 4000 year old pot, believed by one of Sir John Marshall's co-authors to originate from a waterwheel. The pot in question, which is small and is provided with grooves (Fig. 77), was found near a well at Mohenjo-Daro by the Indus, but the well is too narrow to have contained a potgarland. Many of the other pots mentioned in the volume have similar grooves, and these are undoubtedly some form of ornamentation. John Marshall: *Mohenjo-Daro and the Indus Civilization,* Vol. 1, London 1931, p. 318.

Fig. 78

There is, however, another case which is more difficult to reject; this consists of a number of pots found by G. Caton-Thompson and E. W. Gardner: *The Desert Fayum.* London 1934, p. 150. See Fig. 78.

> ... The reservoir was brimful of blown sand; nothing was found before the 9th ft. [2.75 metres] level. .... but lower in the silt down to the bottom, were some quantities of coarse earthenware vessels figured pl. *XCVIII*, 2. These jars, about fifteen in all, are typical *Qawadys* [qādūs, qawādīs (*pl.*). Arabic term for sāqiya pot] serving for attachment to a water-wheel; they appear not to have been previously recorded earlier than the late Greek or Roman period, and their date of introduction into Egypt has been problematic. The present discovery in an early Ptolemaic reservoir is therefore of considerable historical interest, and suggests that the introduction of the sakkiyah into Egypt was of Persian date. ... filled during inundation .... coin of Ptolemy *II* 265 B.C.

This has been commented on by Ugo Monneret de Villard: *La nubia romana.* (Pubblicazioni dell'Istituto per l'Oriente) Roma 1941, p. 44

> Altri hanno sostenuto. ...... Some have held that the qādūs appeared in the Ptolemaic period and perhaps also came from Persia. This is the view of Caton-Thompson, on the basis of a specimen

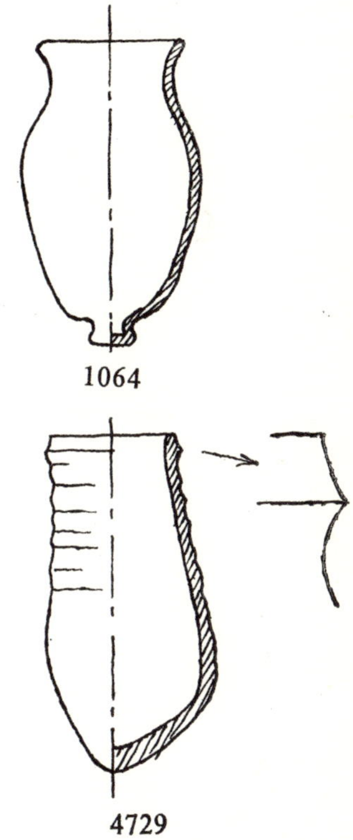

Fig. 75. Roman noria pots from Aleppo. Scale 1:7.5.

Fig. 76. Pot from Israel, 4th to 6th century A.D. Scale 1:5.

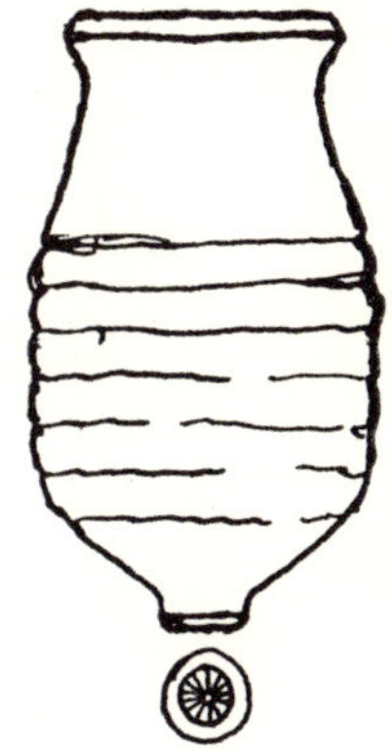

Fig. 77. Mohenjo-Daro pot. Scale 1:5.

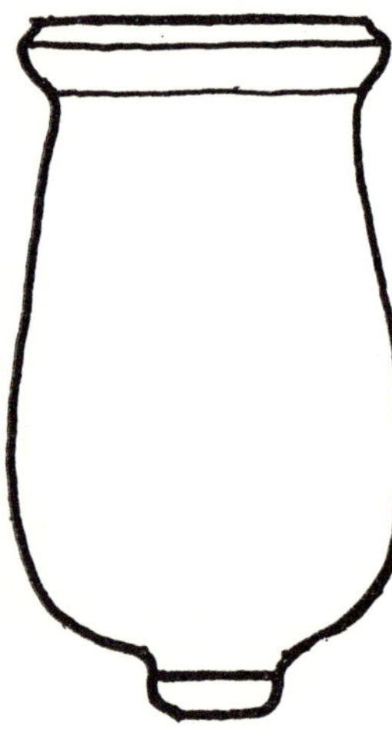

Fig. 78. Pot from about B.C. *265 (?). Scale 1:5.*

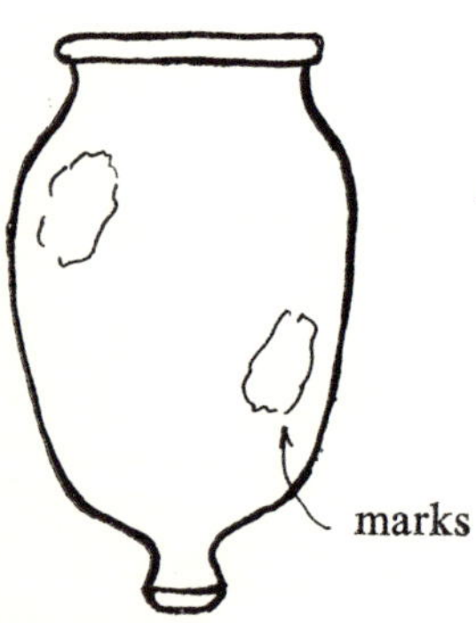

Fig. 79. Pot from the 4th to the 6th century A.D.

found in a water tank in Fayyūm from the Ptolemaic period. However, the tank has probably been out of use for centuries, and the pot could easily be from a later period than the tank.

There is another question that requires consideration. The tank had a capacity of about 300 cubic metres, but Miss Caton-Thompson suggests that there were only about 80 cubic metres in it during the dry season. This small amount of water could be raised by a sāqiya in as little as 15 to 20 hours. Indeed, this tank does not need a sāqiya at all. On the other hand, Miss Caton-Thompson found 15 pots here, lying close together, and this suggests the existence of a potgarland. There is, however, another type of potgarland machine, which is driven by manpower. See Fig. 56, which shows an example from India. Another example, from Egypt, is shown in a drawing in Carsten Niebuhr: *Reisebeschreibung nach Arabien und anderen umliegenden Ländern.* Kopenhagen 1774, plate *XV*, No. *II*. One of the 15 pots found by Miss Caton-Thompson is now in the Department of Egyptology at University College, London, No. U.C. 15578, and we shall now take a closer look at this. When a pot has been used for some months in a sāqiya, it always becomes marked from rubbing against the wheel or the guide-rods, or perhaps even the wall of the well. Such marks can be very deep. I am indebted to my friend, Mr. Hugh Campbell, for having discussed this pot with Dr. A. J. Arkell. They found no marks of this nature on the pot in London, but they did find them on another pot (cf. p. 107).

Ugo Monneret de Villard, p. 45 (cf. p. 101) has many references to qādūs pots and thinks that only the two from Kōm Auṣīm can be dated as they contained papyri from the end of the 3rd or the beginning of the 4th century. Today, these two pots are tucked away in the Egyptian Museum in Cairo, Nos. 55782 and 54696. The former is 326 mm high, and the latter, 353 mm high. I owe this information to the courtesy of Dr. Paul Hönigsberg in Cairo.

Fig. 79 W. M. Flinders Petrie: *Ehnasya.* London 1904–1905, p. 30 and plate *XXX* No. 46: (See Fig. 79).

> ... No. 46 with a knob at the base, and the wide mouth with a brim seems intended to be lashed onto a water-wheel, like modern jars of the same form.

This pot is also in University College, London. In a letter on the subject, Dr. A. J. Arkell writes. "If this is the pot then the drawing in *Ehnasya XXX*, 46 is a very bad drawing." The pot in question is much bigger and heavier than that of Miss Caton-Thompson and has marks from scraping against the wheel or perhaps guide rods. The pot is not dated.

Fig. 80 O. H. Myers and H. W. Fairman: *Excavations at Armant 1929–31.*

Fig. 80. Pots from graves, 4th century A.D. Two of them with bottom air vents. Scale 1:10.

Fig. 81. Pots from graves, 4th century A.D. Scale 1:10.

[The Journal of Egyptian Archaeology, Vol. *XVII* (London 1931), p. 223 ff.].

The first pot in plate *XLV* 205 is 372 millimetres high, but the knob is missing, so it must have been about 400 millimetres high overall. This means that it had a capacity of about 11 to 15 litres, which is rather a large quantity.

The next pot is shown in plate *XLVII*. This holds about 8 litres, *i.e.* the same as the modern pot in Ibiza. However, the point of interest about this pot is that there are four holes in the base. These must be air vents to permit the enclosed air to escape as the pot is submerged. A pot without such vents would float on the water due to the Archimedean upthrust. Unfortunately, none of the other five pots has such holes. If we can rely on the drawing, they are all exactly like that shown in plate *XLVII*, but without the holes. All of them were found in graves. On p. 229 we read:

> A curious feature for which we have not been able to account is the large punctures in the base of certain pots 83 *k*, tomb 753, and 70 *x* 1, tomb 909 [plate *LI*]. In the case of the former, which is a *ḳadūs*, we thought at first that these holes might be due to wear, but with the aid of the proprietor of a saḳia we established the fact that the pot had never been used. [The owner probably noticed that there were no signs of wear.] The holes ranged from four to seven in number and from 1 to 2 centimetres in diameter. Other specimens not published here were found.

Fig. 81

R. Mond and O. H. Myers: *Temples of Armant*. London 1940, plate *LXI*.

No. 72 *A* is rather large and must hold about 13 litres. The next two pots 72 *c* and 72 *c* 1 are rather small, with a capacity of about 1.7 litres. The text to these pots reads as follows:

> ..... Water-wheel pots form class 72. These served other purposes and have been found in Fourth Century graves.

Fig. 82. Pots from A.D. *275–395.*

Fig. 83 a. Late-Roman pots.

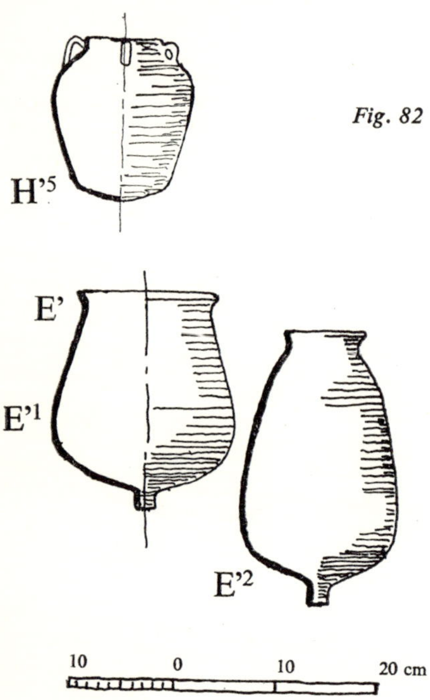

The authors are familiar with this type of pot, and they mention some in the following report.

Fig. 82 R. Mond and O. H. Myers: *The Bucheum*, Vol. *III*. London 1934, plate *CXLIII*.

Here we find four types of pots for the sāqiya, each with a capacity of about 9 litres. Pot *E* 10 is of particular interest because the knob is replaced by a long projection. The pot marked *K* must be similar to that shown in Fig. 80 from O. H. Myers and H. W. Fairman. The pots are dated to the early 4th century A.D.

This chapter on qādūs pots might have commenced with the words of Sir [W. M.] Flinders Petrie: *Wisdom of the Egyptians*, London 1940, p. 143:

> ... The *saqieh* or wheel with band of buckets, was used in Roman times, as proved by the *qādūs* pots found in that period, but not before.

Other pots are mentioned in J. E. Quibell: *Excavations at Saqqara.* (Service des antiquités de l'Égypte, Fouilles à Saqqarah, Cairo):

Vol. *I*, Le Caire 1907, plate 33_1 ...... a knob of a qādūs pot.
Vol. *I*, Le Caire 1907, plate 63 .... a photo of a qādūs pot.
Vol. *II*, Le Caire 1908, plate 63_3 ...... a Coptic qādūs pot.

Fig. 83 a Uvo Hölscher: *Post-Ramessid Remains.* The Excavation of Medinet Habu. Vol. *V* [The University of Chicago, Oriental Institute Publications, Vol. 66 plate 48 and p. 74.]

> ... *E'* 1. Water jug (cadus) with knob base; coarse red clay. Found with *H'* 5 in left sacristy of small Coptic church in squares *B* 7–8. It was let into the stone pavement up to its rim and contained a bowl ....
>
> *E'* 2. Water jug (cadus) knob base; coarser red clay, white wash. Found in square *B* 5. Late Roman.

In some reports on excavations there are drawings or pictures of qādūs pots, but no mention of the term qādūs. So, for example, in J. Schwartz et H. Wild: Fouilles Franco-Suisses. Rapports *I*, *Qaṣr-Qārūn / Dionysias 1948*, Le Caire 1950, plates *XI a* and *e*. These pots come from the western side of the big lake Qārūn, in the Fayyūm.

Details of a number of Coptic qādūs pots are published in H. E. Winlock and W. E. Crum: *The Monastery of Epiphanius at Thebes.* (Publications of the Metropolitan Museum of Art. Egyptian Expedition, Vol. 3, ed. Lythgoe, N.Y. 1926). The pots are shown in plate *XVIII* and there are a few drawings on p. 64. Pots *A* to *F* have a capacity of about 5 litres.

Fig. 83b. Coptic pots.

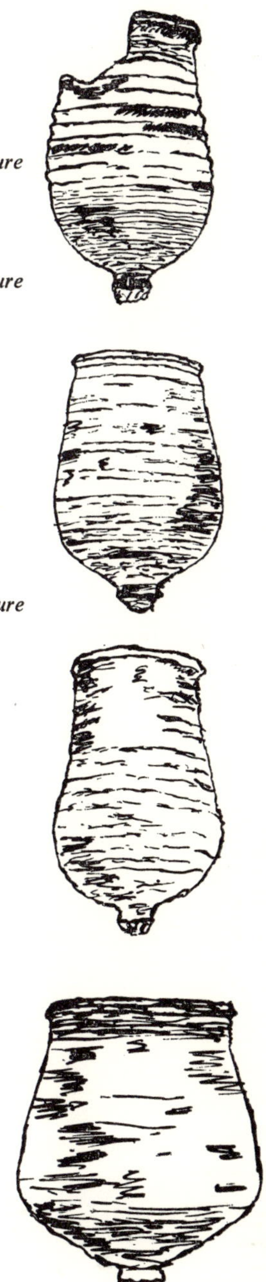

Karl [Carl] Maria Kaufmann: *Die Menasstadt und das Nationalheiligtum* ...., Leipzig 1910, plate 50. Some of these pots are perhaps qādūs pots; a sāqiya has at any rate been in use in this well (cf. p. 136).

No figure

Monneret de Villard (cf. p. 101) writes on p. 45:

No figure

> ... The oldest examples of qādūs found in Nubia are those from Dakka. *Archaeological Survey of Nubia. Report for 1909–10*, p. 36, fig. 6, 28. These types show that we are still at the experimental stage, in which not all parts are completely perfect, as they were finally, when they must have been satisfactory.

Besides the drawings in fig. 6, there is a photograph of this pot in plate 24, *b* 5. The next two references are also taken from Monneret de Villard.

Oric Bates and Dows Dunham: *Excavations at Gammai.* [Harvard African Studies, Vol. *VIII*, Varia Africana *IV*, Cambridge, Mass. 1927]. On p. 63 we read:

No figure

> ... Sākia-pot of hard red ware like Plate *LXII*, fig. 14. Groove worn in one side at [the] neck and another at greatest diameter (through use on Sākia).

On p. 62 and on p. 63 we read:

> ... Grave *E* 99. Sākia-jar of hard red ware, red-washed corrugated. Poorly made and of imperfect shape. Height 34.9 centimetres, diameter 19.2. [See plate *XXII*,1.]

> ... Grave *E* 115. Sākia-pot of hard red ware, wet-smoothed; *tibbin*-marks [marks left on the pot by burning out of chopped straw]. Height 32 centimetres, diameter 20.1. Like Plate *LXII*, fig. 14.

On p. 105 we find:

> ... 11) Of even more frequent occurrence in *E* graves, and not found elsewhere at Gammai, is the hard red ware sākia-pot shown in Plate *LXII*, fig. 14. Ten examples of this vessel were found in *E* graves. It is to be noted that in two cases the pot showed clear signs of

wear on one side below the rim, and between the base and the knob, proof that these vessels were not of special manufacture as tomb furniture, but had seen actual service. This same vessel is noted as belonging to the X-Group in Lower Nubia, and is figured in *The Nubian Archaeological Survey 1907–1908*, Vol. *I*, p. 346, fig. 330,14.

There are good pictures in plate *XXII*, 1 and 3, and in plate *XXIX*, 2 *E*. The period *X-Group* means A.D. 200–600.

No figure Other pots from Nubia are illustrated in F. Ll. Griffith: *Oxford Excavations in Nubia 1910–1913.* [Annals of Archaeology and Anthropology. University of Liverpool, Vol. *XI*, No. 3 (1924)]. Plate *XXI*, 25 a and 25 b, shows two pots from the period A.D. 250–550.

No figure L. P. Kirwan: *The Oxford University Excavations at Firḳa.* Oxford 1939, plate *XXV*, shows another pot, possibly from the 6th century, which was found about 200 kilometres south of Wadi Halfa.

No figure There is also a qādūs pot in Walter B. Emery: *The Royal Tombs of Ballana and Qustul, Mission Archéologique de Nubie 1929–1934.* Cairo 1938, plate 111_{14}.

No figure During the past few years several more pots have been found by the *Scandinavian Joint Expedition to Nubia.* One of these is now in Stavangar Museum in Norway, while another is mentioned by T. Säve-Söderbergh: *Preliminary Report of the Scandinavian Joint Expedition.* [Kush, Journal of the Sudan Antiquities Service, Vol. *X* (Khartoum 1962), p. 104].

Fig. 84 There is a very important paper on qādūs pots in the same volume by

Fig. 84. Dating of pots in Nubia.

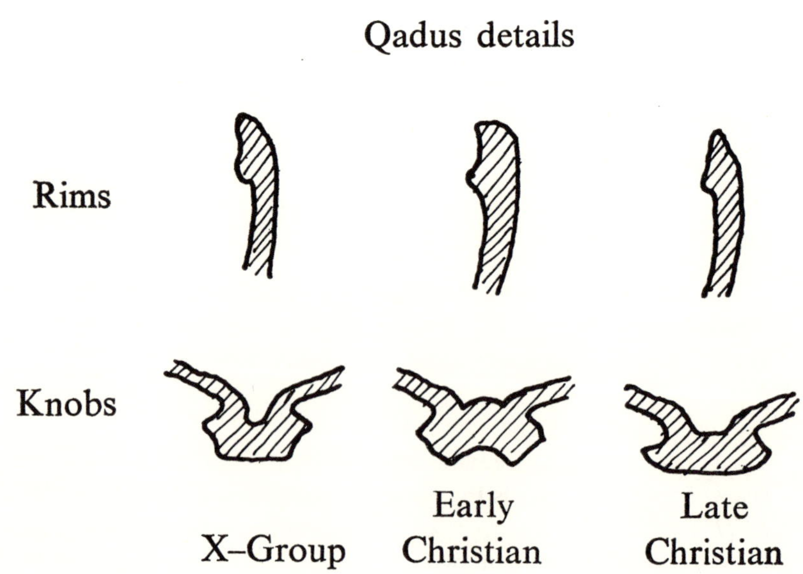

William Y. Adams: *An Introductory Classification of Christian Nubian Pottery*. [Kush, Vol. *X*, p. 251, fig. 7 and 9]. See also [Kush, Vol. *XII*, p. 132].

In this volume Dr. Adams states that the small qādūs pots are the oldest. From this I conclude that the qādūs pot found in Israel (Fig. 76) is perhaps from the first or second century of our own era.

No figure

Thirty-five examples were found at Arminna West by Kent R. Weeks: *The Classic Christian Townsite at Arminna West*. (Publications of the Pennsylvania-Yale Expedition to Egypt No. 3. New Haven and Philadelphia 1967, p. 50).

3. *Unpublished qādūs pots*

At University College, London, there are four qādūs pots, two of which are shown on Figs. 78 and 79. Regarding the other two, Dr. A. J. Arkell has written to me as follows:

> . . . A small one, U.C. 15579, is marked 'KOPTOS' and is probably Roman or later. With marks of wear. The other unidentified one is possibly Ptolemaic.

Fig. 85

The Museum of Agriculture in Cairo (al matḥaf al ʿzirāʿī, ʾal Duqqī) has several specimens of agricultural implements from the Dynastic and Roman periods, including two qādūs pots. One of these, No. 877, is a small type of pot and has 13 grooves. Beside the pot lies the rod to which it was once tied, and there is a little piece of light twine still lashed to the rod. The second pot, No. 878, is bigger and also has a rod. This rod, No. 2493, has been fashioned with great care so as to fit the rim. The rods and pots, which were found in the Fayyūm, are from the Roman period. I have a few, unfortunately rather poor, photographs of these pots.

Fig. 85. Roman pots from the Fayyūm with rod for lashing in place. Scale 1:5.

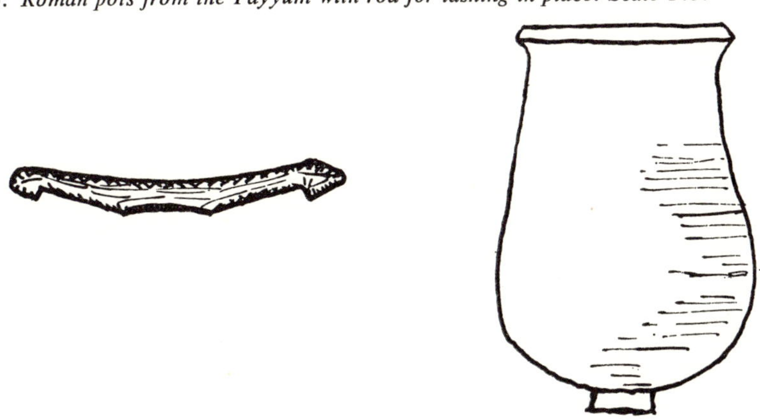

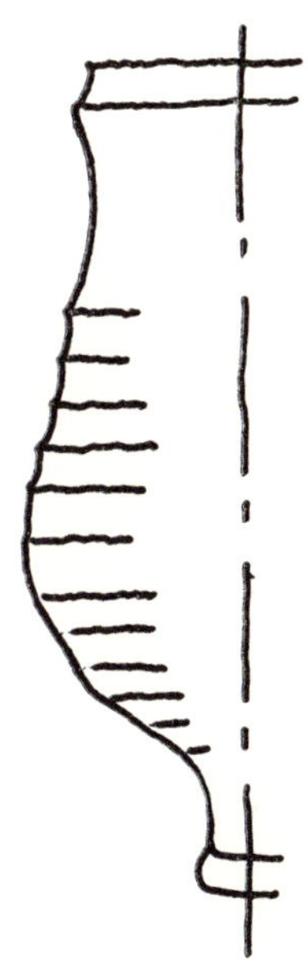

Before concluding, I should like to cite a few words by Abbott Payson Usher: *A History of Mechanical Inventions*, Harvard U.P. 1954, p. 129.

> ... We have scarcely any early drawings, and we know little of the history of the noria before the first century B.C. Winlock's analysis of pottery at Thebes suggests a line of investigation that seems not to have been pursued with much care. (cf. p. 105).

Conclusion

All the pots except those of Miss Caton-Thompson seem to be from a period after the 3rd century A.D. Most of the pots were found in cemeteries, and one was built into the pavement of a Coptic church. In the Sudan, the pots are so common that they seem to indicate that the sāqiya was introduced into the Sudan during the period A.D. 100–300.

William Y. Adams: *Post-Pharaonic Nubia in the Light of Archaeology. I.* [The Journal of Egyptian Archaeology, Vol. 50 (London 1964), p. 119]:

> ... The *causa prima* for the Meroïtic and Ptolemaic renaissance is almost certainly to be found in the introduction of the saqia, which made agricultural productivity for the first time relatively independent of the level of the Nile. So long as pits could be deepened and ropes lengthened, irrigation could be continued even at the lowest period of the annual flow. There can be no doubt at all that the saqia led directly to the reoccupation of Lower Nubia, where steep banks and deep channels had always made irrigation difficult. This theory is supported by the fact that sherds of saqia pots (which are very easily recognized by their fastening-knobs) are invariably found in Meroïtic sites in Lower Nubia.

4. *Objects of wood etc.*

So far, very few objects of wood used in the construction of the sāqiya have been cited, but it is likely that a certain number have been discovered.

The Egyptian Museum in Cairo has a room containing Dynastic and Roman implements of wood, but nothing here seems to have any connexion with the sāqiya.

[W. M.] Flinders Petrie: *Objects of Daily Use.* (British School of Archaeology in Egypt, London 1927, p. 48):

> ... No. 126. Pawl of saqieh wheel of *sont* (acacia) wood. Roman?

There is no drawing or other information. This item may be in University College, London, but there is a strong possibility that it was destroyed by fire during World War *II*.

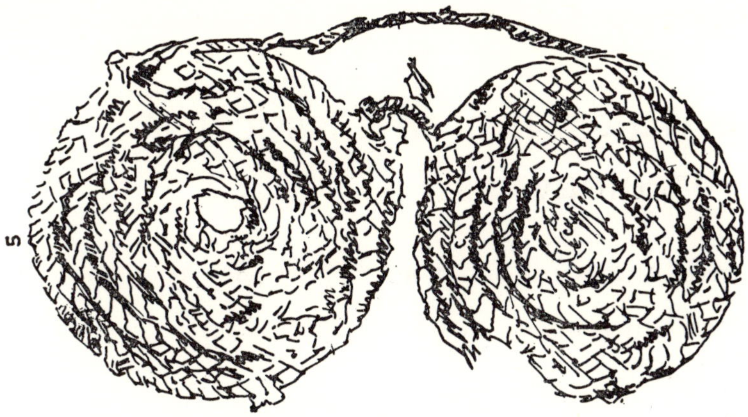

Fig. 86. Blinkers for the sāqiya ox. Found in the Fayyūm.

J. E. Quibell: *Excavations at Saqqara 1908–1909, 1909–1910.* The Monastery of Apa Jeremias. (Service des Antiquités de l'Égypte. Le Caire 1912, p. 29):

. . . a leather bucket from a water-wheel . . .

No other information.

B. P. Grenfell and A. S. Hunt: *Fayūm Towns and their Papyri.* (Egypt Exploration Fund, Graeco-Roman Branch. London 1900). This book gives only a photograph, in plate *XVII*, and the term "blinkers" on p. 47, but the "glasses" probably belong to an ox for a sāqiya. Dating from the 2nd or 3rd century A. D. *Fig. 86*

In the Agricultural Museum in Cairo there are two Roman rods for a qādūs (cf. p. 107).

Vignette No. 6. Abraham at the well. Painting from the synagogue in Dura-Europos. Du Mesnil du Buisson: Les peintures de la synagoguu de Doura-Europos 245–256 après J-C. *Roma 1939, plate I*^{b}.

CHAPTER 7

The Waterwheel in Classical Archaeology

Records and ruins from B.C. 1100 to A.D. 600 reviewed in inverse chronological order

1. PAPYRI

There are several hundred Greek papyri mentioning irrigation, and approximately twenty of these seem to relate to waterwheels. Most of the Greek papyri have been found in excavations carried out in the 19th and 20th centuries. Some of them have been unearthed by archaeologists, others have been chance finds, and the material has now been spread abroad to about 225 museums and libraries all over the world. Gradually – very gradually – the Greek papyri are being published; today we know the texts of about 40,000 papyri and are still awaiting the publication of perhaps as many as 200,000 more.

The Greek papyri have many advantages: firstly, they are original documents; secondly, most of them are based on everyday events; and thirdly, more than half of them can be dated to within a single year, and the remainder to within a century. Unfortunately, only a small number of the documents has been translated, and the translations must be treated with caution. Where the texts contain technical terms, the work of interpretation is very difficult because the dictionaries lack precise translations. However, there are indications that the Greek scribes did not themselves possess a clear terminology; not until the 5th and 6th centuries do we notice a certain degree of system in the use of the different terms.

The selected papyri begin with a Coptic papyrus from the 6th or 7th century A.D. and end with a petition from the 3rd century B.C. At the same time, it will be noted that the interpretations tend to pass from fairly reliable to very unreliable, and this may be due to our passing the time of the introduction of the sāqiya in Egypt. The subject of waterwheels has been dealt with by papyrologists previously, but since then, the material has increased considerably. The papyrologists in question are:

Theodor Reil: *Beiträge zur Kenntnis des Gewerbes im hellenistischen Ägypten*. Leipzig 1913, pp. 82–84.

A. Calderini: *Macchine idrofore secondo i papiri greci.* [Rendiconti Istituto Lombardo di Scienze e Lettere, Rendiconto ser. 2, Vol. 53 (Milano 1920), pp. 620–631]. In the following referred to only as Calderini. Michael Schnebel: *Die Landwirtschaft im hellenistischen Ägypten*, Vol. 1. [Münchener Beiträge zur Papyrusforschung und antiken Rechtsgeschichte, Vol. 7 (München 1925), pp. 74–84].

In papyrology, several hundred abbreviations are used, most of which are given in: *Lexikon der Alten Welt*, ed. Klaus Bartels und Ludwig Huber. Zürich 1965, cols. 3389–3402.

Example of abbreviation:

267 P. Oxy. 1475, 16. The first figure refers to the year. A dated century is given i Roman numerals; thus, for example, a papyrus belonging to the third century is written *III*. This is followed by the international abbreviation, which in the present case refers to The Oxyrhynchus Papyri, which is in more than 40 volumes. This is followed by the papyrus number and, lastly, the line number.

The selected papyri

One of the most important pieces of evidence about the sāqiya is to be found in a Coptic papyrus, which mentions a big wheel with 40 cogs. It is actually this figure of 40 that makes it possible to distinguish this machine from other possible water-raising devices. The sāqiya and the tābūt are the only machines that have a big cogwheel with 40 cogs.

P. Copt. Basil 1 *VI–VII*

W. Spiegelberg: *Vertrag über die Vermietung der Ausrüstung eines Wasserrades.* [Abhandlungen der königlichen Gesellschaft der Wissenschaften zu Göttingen. Philologisch-historische Klasse, Vol. 16 (1917), pp. 77–82]. This publication includes a commentary by U. Wilcken, which should be read first and not overlooked, as is the case in the translation to be found in G. F. Zimmer: *The Chain-of-Pots in the Sixth to the Seventh Century*. [Transaction, The Newcomen Society, Vol. *IV* (London 1924), p. 30]. Translation from German:

> I Papostolos, son of Sarapion – my mother is called Anna – live in Nagos in the nome of the town Shmoun (Hermopolis); I write to the Scribe Phoibammon, son of the Holy Georgios from Shmoun (Hermopolis).
> Behold! there is a big wheel [horizontal wheel] having 40 plugs [cogs] with the potgarland wheel (antlias); this I have hired from thee. I will give thee 19 measures of wheat for this. The measures should be standard measures.

> These I will give thee in the month of Epiphi [25 June to 24 July] in the time of the harvest in the third Indiction, God willing.
> As (soon as) the irrigation is finished I undertake to return it [the wheel] to its proper place in the same condition as when we took it over. If we fail to return it to its proper place, or if someone has stolen it, then I will pay thee 27 keration for it without recourse to law and without doubt.
> I Papostolos, Son of Sarapion, approve this agreement. Written the 28th day of Phaophi [25 October] in the 2nd Indiction. Anuthis, Son of Victor the Deacon, as witness. This was written by me, Victor, the writer, and by God.
> (on the verso) Contract: Papostolos, Son of Sarapion, of the village of Nagos in the nome of Hermopolis.

German translation:

> Ich, Papostolos, der Sohn des Sarapion – meine Mutter heißt Anna – ich aus Nauoje im Gau der Stadt Schmun (Hermopolis), ich schreibe dem Schreiber Phoibammon, dem Sohne des seligen Georgios aus Schmun: Siehe da ist ein großes Wasserrad auf (an) dem 40 Pflöcke mit ihren Schöpfgefäßen (antlias) sind, – die habe ich von dir gemietet. Ich bin bereit dir 19 Maß Weizen für sie zu geben nach dem richtigen Maß. Diese bin ich bereit, dir im Monat Epiphi zu geben von der Ernte der dritten Indiktion, so Gott will. Wenn die Bewässerung eingestellt wird, bin ich bereit, sie wieder an ihren Ort zurückzubringen, wie wir sie übernommen haben, dergestalt, daß wenn wir sie dir nicht an ihren Ort zurückgebracht haben oder man sie uns weggenommen hat, ich bereit bin, dir 27 Keratien für sie zu geben, ohne Gericht und Gesetz, ohne irgend einen Zweifel.
> Ich Papostolos, der Sohn des Sarapion, bin mit dieser Vertragsurkunde einverstanden.
> 2. Hand: "Geschrieben am 28. Phaophi der 2. Indiktion".
> 3. Hand: "Anuthis, Sohn des Victor, des Diakons, ich bezeuge die Sicherheitsurkunde".
> 4. Hand: "Durch mich Victor, den Vertragschreiber, wurde es mit Gottes Hülfe geschrieben".
> Rückseite – 5. Hand: "Sicherheitsurkunde ..... Papostolos, Sohn des Sarapion, aus dem Dorfe Nagos im Gau von Hermopolis".

The first of Spiegelberg's problems concerns *the big wheel.* He believes this to be a potgarland wheel, but U. Wilcken writes: "... erlaube ich mir, eine abweichende Deutung von *wheel* (hoi) zur Prüfung vorzulegen." The *wheel* must be the horizontal wheel, firstly because it has 40 cogs, and secondly because of another text, which refers to a boy sitting on the *wheel.* The boy would obviously not be sitting on the potgarland wheel.

The second problem relates to the *antlias.* This is probably the potgarland wheel, but Spiegelberg sees it as the potgarland. Later on (p. 81), he says that the text should be interpreted to mean that only the potgarland was rented. This I find unlikely. Today it is still possible to rent the whole machine, as described by M. H. Davidsen: *Among the Fellahen of Upper Egypt.* [Geografisk Tidsskrift, Vol. 49 (København 1948–1949), p. 25 ff.]. The rent of a sāqiya is also mentioned by W. Nicholls: *The Sakia in Dongola Province.* [Sudan Notes and Records, Vol. 1 (Khartoum 1918), p. 23]:

> The amount paid for the hire of the waterwheel varies sometimes, as, if there is a shortage of wheels and a good demand for them, the owners of them naturally put up their prices.

Spiegelberg has yet another problem to tussle with in the fact that the term for the machinery sometimes appears to be used for the irrigated land. This double-duty is fairly common in the Greek papyri and is still applied today. In the Sudan, the term sāqiya is used for the land, and the old Persian term dawlāb, for the machine. See H. A. Nicholson: *Saqiya Terminology in Dongola.* [Sudan Notes and Records, Vol. 18, Part 2 (Khartoum 1935), p. 314].

Spiegelberg has references to other Coptic texts. Some of these and others are to be found in W. E. Crum: *A Coptic Dictionary*, Oxford 1929 p. 650 b (hoi), but since then, at least two other Coptic sāqiya-texts have appeared. W. E. Crum and H. G. E. White: *The Monastery of Epiphanius at Thebes*, Part II (The Metropolitan Museum of Art, Egyptian Expedition, N.Y. 1926, p. 237):

> Ostracon No. 312
> ..... thou give him the pots for the water-machines.

The Coptic word for machine is derived from the Greek term *organon.*

CPR IV 133 (Corpus Papyrorum Raineri) *VII*
Walter C. Till: *Die koptischen Rechtsurkunden der Papyrussammlung der Österreichischen Nationalbibliothek.* Wien 1958, pp. 125–126:

> ... my water-machine which irrigates for you ...

A large number of papyri have been identified as sāqiya-papyri because the text contains the term *axōn*, translated "shaft for machine". However, I believe that *axōn* implies more than a shaft for a wheel; in fact, there is much to indicate that it should be translated "the vertical suspended rod of a šādūf". See Fig. 87.

P. Oxy 2244, line 34: *VI–VII*

> To Johannes, the son of Paleus, from the village of Leon, who is employed as a farm worker on the machine/field (mēchanē) at the

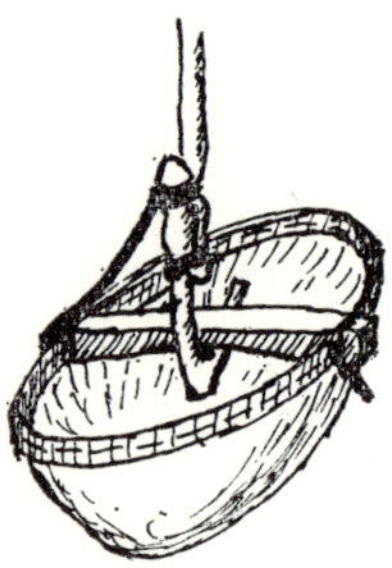

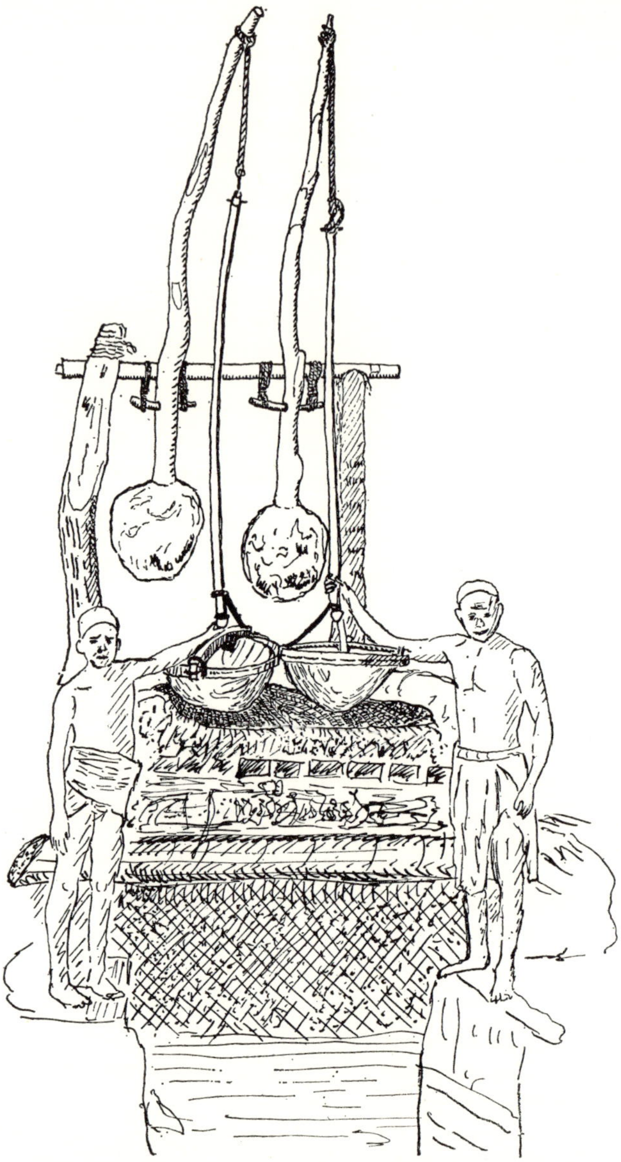

Fig. 87. Egyptian šādūf. This consists of a leather bag fastened to a wooden ring. The ring is lashed to a crossbar, and this is connected to the šādūf-rod by means of a hook. A short rope fastens the šādūf-rod to the lever arm. This turns upon an axis which is suspended from a beam.

The work lies in pulling the leather bag down to the water, whereas lifting it is a relatively easy matter. The work cycle has a duration of 7 to 8 seconds, and each time, about 10 kilograms of water are raised a height of 2 metres. Maximum output is five cubic metres per hour per man. This is 28 watts.

new well (?) (lakkos), the field (mēchanē) called Limenion [followed by the date]: 1 hand-shaft. He does not possess an old one.

What do these words conceal? We are at the house of the steward who supplies šādūf-rods – not sāqiya-shafts – to the farm workers. A sāqiya-shaft, which weighs more than 100 kilograms, would hardly be called a hand-shaft or, as the Oxyrhynchus editor writes, "worked by hand".

Line 41:

To Johannes, son of Phoibammon, from the village, Panguleion,

> farm worker employed at the field (mēchanē) called Tekon, the 25. Tybi of the 6th Indiction, of those [šādūf-rods] received from Alexandria: 1 šādūf-rod (axōn).

The estate in question is located two to three hundred kilometres south of Alexandria. If the parts mentioned really were sāqiya-shafts, they would hardly have been obtained from Alexandria because wood from this city probably means imported timber from the Lebanon, whereas the sāqiya-shaft was always made of acacia wood, and the acacia tree is indigenous to Egypt. Here we may mention a papyrus concerning the acacia tree:

P. Rendel Harris 80, line 18 ff.: *250*

> ... Please supply for cutting from the acacia trees standing by the irrigators (hydreumata) [sāqiya (?)] one acacia tree for making a rim (?) (hapsis). ... The nails will come from [the machine] which is being refitted.

The interpretation of *P. Oxy. 2244* and many other agricultural papyri is difficult because *mēchanē* can mean both machine (šādūf or sāqiya) and a mechanically irrigated field. In the next text, the term must mean machine, *i.e.* šādūf or sāqiya.

> Line 45 ff.: (*P. Oxy. 2244*)
>
> To the superintendent, Joseph, and the bath attendant, Joseph, at the baths in the village of Panguleion, for use in [the šādūf] (mēchanē) at the baths in question ..... 1 [šādūf-rod] (axōn).

In line 34 f., the text ends: "He does not possess an old one [šādūf-rod]." From *P. Oxy. 1911* and *P. Oxy. 137*, we know that a šādūf-rod was normally used for about seven years, and we are also informed that some of the rods "are supplied new" (... of those brought now) *P. Oxy. 2244*, 24, 56, 58. These rods have to be long and free of knots. Today, a šādūf-man never leaves his machine without taking the rod with him because it is a costly implement.

The part of the papyrus that has been preserved mentions 51 *axōn*s, and three of these seem to fit the dictionary translation, "shaft for sāqiya". In lines 83 and 85 we read: "*axōn*s supplied by the church." What is important, however, is the fact that the *axōn*s are to be used for the *kykleutērion*s. According to the dictionaries, *kykleutērion* means circumference, waterwheel, *i.e.* sāqiya. The translation may be correct because *kykleutērion* is normally attached to a *lakkos*, which undoubtedly means: well. If the dictionary translation is correct, then *axōn* must mean shaft for the sāqiya.

In the exceedingly important papyrus, *255 S.B. 9408–9409*, it is clear that in one place, *axōn* stands for sāqiya-shaft, and in another, for šādūf-rod. For the sake of maintaining chronological order, this will be discussed later (cf. p. 122, (7)).

A large group of sāqiya-papyri relate to repair contracts. These are almost identical, whether they are from the 3rd or the 6th–7th century, and their layout is usually as follows:

After an introduction, the name of the field (mēchanē) in which the machine is located is mentioned. This is followed by the spare part needed, which is often an *axōn*, although we sometimes encounter a term that has no precise translation. The spare part is made by the carpenter, but he must first receive a requisition from the actual owner of the machine. The contract ends with a passage stating that the spare part must be in order or that it must fit the old machinery.

In 1913, Reil (cf. p. 110) introduced in papyrology a translation of a technical term which remained in use up to 1965, when F. T. van Straten proposed another interpretation. The term in question is *ergatēs*. Reil translates this as *die Winde*, in the sense: *das Gangspill, die Ankerwinde, i.e.* – capstan. However, from a study of a picture – incidentally, rather a poor one – of a sāqiya, van Straten finds it reasonable to translate *megas ergatēs* as "a big cog-wheel" and *mikros ergatēs* as *die davon ausgehende Stange*. If van Straten had had a better photograph, he would undoubtedly have translated *mikros ergatēs* as "the small cog-wheel"; instead, he cautiously uses the term, *ein kleiner Bestandteil*.

VI–VII *P. L. Bat. XIII No. 20.*

Übersetzung:

..... Da auch jetzt für das in unserer Benutzung stehende Wasserwerk (field) des Landherrn, welches "Das südliche" genannt wird, und das für das Weinland und das Bauland Wasser schöpft, Mangel besteht an einem kleinen Bestandteil [a small cog-wheel], sind wir in die Stadt gegangen und haben wir Eure Berühmtheit gebeten Auftrag zu erteilen, daß uns der erwähnte kleine Bestandteil [small cog-wheel] verschafft würde, und sofort hat Eure Berühmtheit, Sorge tragend um die Instandsetzung ihrer Sachen, uns diesen verschaffen lassen durch Pateuos, den dortigen Maschinenbauer, einen neuen Bestandteil [cog-wheel], passend, geeignet für ein Wasserwerk (antlētikon) [potgarland wheel] und gediegen. Wir haben ihn empfangen zur völligen Wiederinstandsetzung des Wasserwerkes (mēchanikon organon) [sāqiya] am heutigen Tag. ....

Contracts dealing with the leasing of fields and machines, or a house with wooden doors, are frequent, and they are sometimes very detailed, as in the following example:

574 *P. Berol.* [P. Berl. Zilliacus No. 7]:

> I offer voluntarily to lease – with effect from sowing-time – ..... the field/machine belonging to your Honour and called Akeeis, which lies to the north of the village. With all wooden attachments and iron fittings [*e.g.* bearing gudgeons and shaft rings]. Included are the appurtenant well (lakkos) and the potgarland (hydreuma) and potgarland wheel (kykleutērion) and the machine [angle gear]. In addition, the appurtenant tools (?) and all other equipment rightfully belonging thereto; and the artificially irrigated field (mēchanē) of 12 *arouras*.

This translation is not in exact accordance with the German; I have changed some of the technical terms. It is noteworthy that the editor translates *lakkos* as a well, whereas most other interpreters take the word to mean a water tank or reservoir.

There is a unique sāqiya-papyrus, which, as evidence of the presence of the sāqiya in Egypt, renders all the other papyri superfluous. This papyrus, which is published in Aegyptus Vol. 37 (1957), p. 93, mentions not only a big wheel but a big wheel with 36 cogs. The wording is almost the same as in *P. L. Bat. XIII 20*, and the passage in question relates to the following:

A peasant asks the landowner whether arrangements can be made without delay for him to get a big new cog-wheel for his machine, which irrigates his vineyard and farmland. The wheel should have 36 cogs and should be made to fit the old machinery. The peasant promises to pay the tax as soon as he gets the wheel this month – September. The document also acts as a receipt for the wheel. There is no copy of the document, and he is conversant with its contents.

S. B. 9503 (SBGU 9503). [Aegyptus Vol. 37 p. 93]. *440–441*

> .... As I now need a big "mechanical" wheel (megas ergatēs mēchanikos) for the imperial field/machine (despotikē mēchanē) called The New One, which raises water for the vineyard and farmland, when I came to the city, I asked that this be procured for me and that Your Honour would supply this new suitable cog-wheel (ergatēs kainos epitēdeios) with 36 cogs (skytalē) ... [See Fig. 88] ..... in this the month of Choiak [September] of the year of .... in order to complete the mechanism of the machine. ..... [and hereafter] I acknowledge that ....

There are texts that make no mention of machines but that still, as in the case of the next papyrus, appear to relate to a sāqiya.

P. Oxy. 938. *III–IV*

Editor's translation:

> .... although you had long been instructed to send twelve baskets

Fig. 88. Egyptian sāqiya. A pad is arranged over the shoulders of the camel to take one end of a long yoke. The other end of the yoke lies in the angle formed by the upright shaft and the inner end of the drawbar. The yoke must not be fixed to the shaft as this would hinder the rocking walk of the camel.

> of hay, with the result that the oxen are in danger of destruction. Since the oxen thus are in a sorry state, and the land in consequence is not being irrigated, I hasten to write to you now once more and beg instantly to get the baskets properly laden and sent off, for you seem to be mocking my industry. I pray for your long health.

The reason why the sāqiya and the noria have been able to compete with modern centrifugal pumps is to be sought in the fact that waterwheels can be repaired by the local carpenter, whereas a centrifugal pump calls for a service station with mechanics and spare parts in the locality. I have myself seen waterwheels working merrily beside idle, rusty motor pumps. The engine was hardly worth repairing, but as the fields had to be irrigated each day, the peasants had rebuilt the old waterwheel.

Nevertheless, even the local carpenter may present problems, as will be seen from a papyrus from the 3rd century.

254 *S. B. 9053*, 8–11 (SBGU 9053).

> .... Please lose no time in sending the machine near Moithymis because we need it.

Among the Greek papyri there are a couple of estate accounts that provide a very vivid picture of the work on a large estate. Although the accounts only contain about ten lines relating directly to waterwheels, these lines lose some of their value if the rest of the text is omitted. One of the best accounts scrolls is at the University Library in Prague and has been published in three volumes of the Czecholslovakian periodical, Listy Filologické, Vols. 80–82.

S. B. 9406 (SBGU 9406).

About *250*

A German translation is available in: Ladislav Varcl: *Aus den Archiven von Theadelphia.* [Listy Filologické, Vol. 80, with the supplementary part, Eunomia *I*, (1957) p. 61].

Line:

3	To myself [the clerk] Eirenaios	40	drachmas
8	To the ox-driver	4	–
9	To the wood-guard	4	–
12	To the donkey-driver Philon	4	–
16	Other workers [probably freemen]		
21	To the ox-driver	8	–
25	To the ox-herdsman	8	–
28	To the donkey-driver	16	–
39	To cost of this papyrus scroll	4	–
40–44	From the town of Arsinoë: 10 *minas* iron, including three ploughshares at 2 dr. per *mina*, total	20	–
45–47	To the carpenter Heron for a bench, a plough and two beams for the oil [press]. 6 days' hewing	24	–

Classical literature contains many references to oil presses, but none at all to waterwheels with right-angle gear, and this is presumably why our knowledge of the former is so extensive, and our knowledge of the latter so limited. For comprehensive accounts of oil presses and mills, reference is made to:

A. G. Drachmann: *Ancient Oil Mills and Presses.* Copenhagen 1932.
H. Camps-Fabrer: *L'Olivier et l'huile dans l'Afrique romaine.* Alger 1953.
Georges Tchalenko: *Villages antiques de la Syrie du nord . . .* Vol. *I.* Paris 1953, pp. 360–372.

After this slight digression we will return to the translation of the German interpretation.

Line:

67	Two men shaking [winnowing], 6 days, 3 loaves a day	36	loaves
74	Donkey-driver from Theadelphia	27	–
76	Heron the carpenter, who made a bench and a plough in 6 days	18	–
81	Two camel-drivers who brought the wood	12	–
146	Oxen for ploughing from 1st to 30th daily, 1 *metron*	9	*artabes*

262	Raising water for irrigating the field	10	(working days)
266	Isidores was ill	30	–
269	Mixing olives (?) [in a] container	18	–
271	List of tools		
272	15 chisels		
273	3 shovels		
274	5 small axes		
275	1 scythe for mowing hay		
276	1 ploughshare for a full-grown ox-team		
277	3 yokes for oxen (?)		
278	3 empty vats		
300	The 2-drachma tax	1000	drachmas
301	Soldiers for palm-fibres (?)	30	–
302	A soldier for clothing	200	–
303	Guards	80	–
304	Messengers	42	–
305	The police inspector	60	–
306	Doorkeepers	8	–
308	The interpreter	8	–

About *246* *S. B. 9408.* (1)

A translation is given in the same volume as *S. B. 9406*, p. 70.

Line:

5	.... for the month of Toth [*i.e.* September, when the inundation starts]	
11	From Sabeinos [an Italian name] lessee of a mill, paid through Aunes	200 drachmas

Here I must digress once more. Very few hour-glass shaped grain mills have been published, apart from those at Pompeii. Others are to be seen at the Museum of Aleppo, the Museum of Ḥamā, the Museum at Palmyra, the Museum of Damascus, the Museum of ʿAmmān and the Museum of Ǧaraš.

Line:

15	Throwing sandbags (?) on the field and dike, 38 [working days] at 2 dr. per day	76 drachmas
16	Throwing sandbags (?) along the field 18 *arouras*, 25 [working days] at 2 dr. per day	50 –
21	Gleaning, 18 [working days] at 2 dr. per day	36 –
24	Fallow ploughing the field of Puaris 20 [working days] at 2 dr. per day	40 –
28	Digging stones out of the well (?) in the goldsmith's field 104 [working days] at 2 dr. 2 ob. per day	237 dr. 5 ob.

About *255* *S. B. 9408.* (2)

A translation is given in the same volume as *S. B. 9408*, (1) p. 79.

Line:

4	The month of Pauni [*i.e.* June, when the Nile is at its lowest]	
52	Cost of 800 pots (keramides) at 8 dr. 4 ob. per 100	68 dr. 4 ob.
53	Cost of 2 [volumes] *gloiaphion* [some sort of adhesive – glue] from Oxyrhynchos	20 drachmas

The word, *keramides*, is normally translated as earthenware pots, and there is a papyrus in which *keramides* are closely connected with *mēchanē* [sāqiya] *e.g. II S. B. 9379*, Col. II, 6 "... payment for pots for the machine 8 *arouras*: 24 dr.".

The question is whether 800 *keramides* is not a very large number. It is indeed! Even in the case of small pots, it represents several wagon-loads. The strange thing is that a similar figure is mentioned in another important papyrus, *113 P. Lond. 1177*, which also concerns machines. This British Museum scroll states that the *keramides* are lashed in position with thin strings. See Fig. 66. Here, it appears that they are glued into position. All the quantities mentioned are big, round figures, from 1600 down to 200, and the price is always per 100.

Keramides may mean pots for the sāqiya, although not everything points to this. The editor of *P. Lond. 1177*, p. 185, in the note to line 158, thinks that *keramides* must be pots because he considers it unlikely that there should have been such a heavy demand for bricks. I personally am inclined to draw the opposite conclusion.

Line:

66	Cost of 20 rings (krikotōn) supplied by Horeion the carpenter and sent to the town by order of Ninnos	120 drachmas
69–71	Purchased for the machine from Hermas the carpenter, sent to the town for ...... by order of Ninnos: five and a half ells long [about 3 metres]	140 –
72–74	Purchase from the same Hermas of two complete ploughs, sent to the town by order of Ninnos	80 –
75–77	Purchased from same [Hermas]: another plough given to the donkey-driver Dorion	24 –
78	10,000 ..... hay at 8 dr. per 100 [feed for the sāqiya-oxen]	800 –
81	Digging a well (poiountes pēgēn) for the oxen	17 dr. 1 ob.
82–83	To Kanopos the carpenter for dismantling the sāqiya (piptōn tympanon) at the goldsmith's field (4 working days) at 2 dr. per day (cf. line 28)	8 drachmas
86	To same [Kanopos] for making a new shaft (axōn) for the estate "The Pumpkin", which took 5 days	20 –
89	Purchase of 4 vats given to the donkey-driver Kastor	16 –
91	Purchase of 4 horsehair [switches (?)] at 52 dr. each	208 –
104	Total expenditure: 3400 drachmas	
	Against receipts: 3076 –	
	Outstanding: 324 –	

S. B. 9409

255 (1) This part of the accounts has not been translated, but the contents do not deviate significantly from the previous part. We are in the month of April when, as the Nile is low, the rushes have to be irrigated. A house with doors is to be built, and once more 1600 *keramides* with adhesive are needed. The carpenter Hermas is again kept busy on the machine.

(5) Kanopos the carpenter needs wood, and this time only 800 *keramides* are needed, together with a correspondingly smaller quantity of adhesive. Unfortunately, the next section has not been translated by Varcl, but the content is as follows:

(6)

To same [carpenter] a new machine (tympanon) for .... 4 working days	16 drachmas
Mounting (?) a new shaft (axōn), 2 working days	8 –
Mounting a cog-wheel (ergatēs) at the Simelius estate, 2 working days	8 –
Shaping pins [*i.e.* cogs] for the machine (skytalas eis tympanon)	

I think I detect a sāqiya behind these lines because of the mention of a *tympanon* and a new shaft, and a cog-wheel with cogs. Further on in the accounts, adhesive is bought again, although this time without *keramides*. The accounts in this section apply to the time immediately before the period of lowest water level of the Nile.

(7) The next section concerns irrigation. Line 93 is a heading concerning 8 shafts. One shaft is despatched to an estate and the other 7 are left over. The next two lines deal with leather and leather sacks. Here, I believe, shafts (axōn) means šādūf-rods, and the leather is for the šādūf-bailer. See Fig. 87. The following lines concern tools.

About *250* *S. B. 9410.*

(5) The first two lines contain a specification of work in the vineyard, after which the editor believes that the expression, "which was sent to the town", should be taken to mean "was made in the town". Right in the middle of the account there is a personal item: "Payment for a little pig for my father." (lines 17–18).

(7) Wood for the machine and rings.

252 *P. Ups. 23.* [P. Berl. Leihgabe 23]

For translation, see Johnson: *Roman Egypt*, pp. 143 and 104 (cf. p. 124):

Sub-lease of $2\frac{1}{2}$ *arouras* of vineyard .... pays for a new waterwheel and provides animal for its operation. Lessee repairs old

> water-wheel ..... the cost of feeding the oxen during the irrigation season.

What makes this papyrus of such great value is the fact that it states that the machine is to be provided with animals and that these are to be supplied with fodder during the irrigation period. The text must therefore refer to a sāqiya.

P. Flor 16. 239

For translation, see Johnson: *Roman Egypt*, p. 143 (cf. p. 124):

> Lease of a garden.
> I wish to lease, for a term of 3 years from the present month of Mechir in the second year, near the village of Euhemeria, a vegetable garden of 1 *aroura* [0.68 acres] or whatever area it may have [a new survey may perhaps show another value] in which is a cistern and water-wheel at a total rental of 100 drachmas per year, subject to no deductions or risks. .... For irrigating the land I have taken over from the noble lady a cow valued at 100 drachmas. Whatever is needed for new equipment, supplies or lumber, nails, stone, pitch, carpenter's wages, transport, water-wheel, yokes, and all state charges shall be borne by her ladyship, while I, the lessee, shall perform the work on dykes and .....

Once more it is the mention of oxen that points towards a sāqiya, but the part about the carpenter also provides an important piece of evidence.

P. Lond. 1177. 113

We tend to think of waterwheels in connexion with agriculture as machines for irrigation, but, in reality, they are used anywhere where water has to be raised, for example, for the supply of water to fortifications and public baths and, as will be seen from the next contribution, for urban water supplies.

The town in question is undoubtedly Arsinoë (Madīnat al Fayyūm) which was situated north of the present town in antiquity. Near the town there was a quay called Alsos, which had a water tower and two sāqiyas operated by 6 ox-drivers. There were also 16 šādūfs at this water tower. When the water level in the canal was particularly low, the sāqiyas and the šādūfs were supplemented by water screws. At another location, called Telesos, ox-drivers toiled all year round, aided by labourers who worked the treadwheel. Here, too, there was a water tower from which the water was led to three of the town's public facilities, a bathhouse, a brewery and a synagogue.

The accounts commence in the month in which the Nile is at its lowest level and end when the inundation is at its peak. The text shows that the machines, which also included the šādūfs, were under constant repair, and we can see that it also deals with spare parts. However, it is impossible to give a precise translation of the various parts, and an incorrect translation would be worse than leaving the question open.

The text has been made the subject of a study by Michael Schnebel: *Die Landwirtschaft im hellenistischen Ägypten*, Vol. 1. [Münchener Beiträge zur Papyrusforschung und antiken Rechtsgeschichte, Vol. 7 (München 1925), p. 74 ff.], and has been partly translated by Allan Chester Johnson: *Roman Egypt to the Reign of Diocletian.* (An Economic Survey of Ancient Rome, ed. Tenney Frank. Baltimore 1936, p. 685). There is a translation in A. S. Hunt and C. C. Edgar: *Select Papyri* ... Vol. *II.* No. 406 (The Loeb Classical Library).

These are actually more in the nature of interpretations than translations, and the problems involved are far more complicated than is implied. I have spent many hours discussing the texts with the papyrologist, Tage Larsen, and with Dr. A. G. Drachmann. These talks have clarified in my mind the terms that are certain and those that are based on a single occurrence or that simply have no basis at all.

Selected texts from *P. Lond. 1177.*

Line:

10 Account of revenues and expenditure for the water supply.

65 Expenditure in connexion with the water tower Alsos, with 16 šādūfs, each with $1\frac{1}{2}$ men, who worked the machine from dawn to dusk.
[Each with $1\frac{1}{2}$ men means that there are 16 šādūfs and 32 men who are paid the equivalent wage of 24 foremen].

72 Payment for lamp oil and lamps for those on nightwork: 12 dr. 2 ob.
[The pay account shows that men worked at night at both the šādūfs and the sāqiyas].

113 Wages for ox-drivers at the water tower Alsos, with 2 machines [sāqiyas] (and) 6 ox-drivers.

140 ... for the days from 18th to 30th of (the) same (month) Pauni, (total) 13 days, on one of which (only) one machine with 3 ox-drivers (was used, while on the) remaining 12 days, 6 ox-drivers (were used), making 72 ox-drivers.
Total: 75 ox-drivers [*i.e.* 75 ox-driver days].

It is these lines (113 ff. and 140 ff.), of the 335 lines comprising the document, that best reveal that the machines in question are sāqiyas.

U. Wilcken has also studied this papyrus, and he concludes that "This papyrus should be studied jointly by a papyrologist and an engineer." [Archiv für Papyrusforschung .... Vol. 4, p. 554.] As there are still at least twenty terms lacking a precise translation, there is plenty of work for such a team.

Fig. 89. The thresher sledge (nawrağ). Fonds Arabe 2964 folio 22 (detail), Bibliothèque nationale. D. T. Rice: Islamic Art. *London 1965 facing p. 100.*

P. Lond. 131 recto et verso. *79*

This papyrus has been translated by Johnson: *Roman Egypt*, pp. 177–207, and p. 197 of this has the following about a waterwheel:

> ... To the bricklayer laying brick for the waterwheel ...

This translation can hardly be correct, for comparing it with the text, *P. Lond. 131* (Vol. *I*, p. 178, line 303), we find the term *kēlōnos*, which the editor thinks should be read as *kēlōneion.* This is always translated as šādūf. It is also found in Hebrew as *kilōn*, see Dalman (1932), p. 224 (cf. p. 41). Incidentally, Johnson himself, p. 201, translates this as šādūf in another passage: ".... repairing the šādūf at Tomis ...".

There is undoubtedly a šādūf in the text, but the question is whether the account refers to a sāqiya as well. As the papyrus in question is one of the oldest documents available on water-raising machines, it should be studied with great care.

The text shows that the machine is operated by different labourers. For several consecutive days, a couple of men make ropes for the machine. The machine is often in use during harvest time; this does not accord well with other papyri, in which the lease of the sāqiyas comes to an end at precisely this time, and the rent is paid.

The question is whether the text can possibly relate to another type of machine altogether. In Egypt, farmers used – and indeed still do use – a thresher (nawrağ), which is equipped with eleven small wheels that roll over the threshing floor. The thresher is pulled by a pair of oxen, whose path describes a circle of about 10 metres diameter. See Fig. 89.

F. L. Smith: *The Egyptian "Norag" and its Origin.* [Bulletin de la Société sultanieh de géographie, Vol. *X*. (Le Caire 1921), pp. 261–269].

In connexion with the machine under consideration, use is made of the term, *kyklas*. A number of the papyri use the compound, *kyllē kyklas*, which denotes a certain part of the sāqiya. Thus in *552 P. Lond. 776* (Vol. *III*, p. 278) and *552 P. Oxy. 197, 476 P. Oxy. 1899, 543 P. Oxy. 1985, 616 P. Oxy. 1991, VI–VII P. Oxy. 224, VII P. Baden 95, 566 P. Michael 43.* Many papyrologists have commented on this compound, but the correct solution to the problem is undoubtedly that suggested by Ugo Monneret de Villard: *La Nubia romana*, Roma 1941, p. 43. *Kyllē* is a derivative of the Arabic qullah, which means pot. If we expand this interpretation, the compound, *kyllē kyklas*, must mean potgarland wheel.

According to Calderini, p. 625, this is the only papyrus in which *kyklas* appears, but it becomes common in the 6th and 7th century records, where it is certain that it relates to the sāqiya.

Bearing this in mind, we cannot exclude the possibility that *kyklas* in this papyrus from the first century A.D. means a thresher. However, there are other factors that indicate the presence of a sāqiya, including the mention of the cleaning of a well and of a channel from the northern water tank. Johnson: *Roman Egypt*, p. 207. This may be the oldest papyrus mentioning a sāqiya, and in that case it is a very poor proof indeed.

B.C. *5* *P. Berol = BGU 1120.*

There are innumerable references to this papyrus because it is accepted as the first papyrus to mention a sāqiya or some kind of water-raising wheel (trochos). The subject is the lease of garden land, including a šādūf and a wheel, and the passage ends as follows:

> . . . at the end of the term they (the lessees) shall hand over the leased properties fully planted and in flourishing condition, also the doors and the šādūf . . . . .

If *trochos* were a sāqiya we should expect to find a few words specifying the condition in which it was to be returned. But only the šādūf and the valuable wooden doors of the house are mentioned. There are many leases in which precisely this question of the return of a sāqiya is dealt with in detail. In this papyrus I believe that *trochos* is identical with a pulley hoist. The full wording of the text is given by Johnson: *Roman Egypt*, pp. 134–135.

III B.C. *S. B. 9302 = (Edfu VIII).*

The papyrus has been translated by Georges Manteuffel: *Quelques textes d'Edfou.* [The Journal of Juristic Papyrology, Vol. 3 (Warsaw 1949), pp. 103–105]. The petition – for that is what the document is – was found at Edfu, 800 kilometres south of Alexandria, in 1939, and is today in Warsaw. A palaeographic investigation has shown that it was written at the end of the 3rd century B.C., and its subject matter is as follows:

The petition is addressed to one of the Ptolemies, and after a short greeting, the writer immediately gives an account of the drought that has befallen the country:

> But, Your Majesty, I will show thee a machine (mēchanē) – from which thou will suffer no loss – which will deliver the country as the river has not risen for the last three years. The drought will cause such starvation that I must eat ... [here there is unfortunately a lacuna so we are left to guess what the poor man must eat]. ... but if thou will (listen to my advice), Your Highness, we shall have a bountiful year.
> I therefore pray thee – if it please thee – to bid the strategus Ariston to take the necessary steps for my requirements for 30 days, and to send me to thee [in Alexandria] as quickly as possible with a view to a meeting, if it so pleases Your Majesty .... May you be happy.

This papyrus has been studied by many scholars, but they have not been concerned with its technical aspect, although it is clear from the text that a new invention is at hand.

Conclusion

Of the two dozen selected papyri, it is particularly *SB 9503* that attracts our attention because it mentions a big wheel with 36 cogs. This papyrus can be dated to the year 440, but the very interesting accounts scroll from the Fayyūm *SB 9406 – 9408 – 9409*, also points to a sāqiya because it mentions that the carpenter mounts a new shaft and a new cog-wheel, for which he shapes cogs. This papyrus goes back to the period about the year 250. Both these papyri were published in 1957 and have therefore not been reported as frequently as the famous papyrus, *P. Lond. 1177*, edited in 1907. Here, it is really only two lines describing two machines with six ox-drivers that hint at a sāqiya, but the staff of craftsmen listed shows us that it must have been a complicated piece of machinery.

We now jump back 400 years in time, and here make acquaintance with a papyrus that shows clearly the interest of the period in irrigation schemes, although we are given no details of the actual machine.

As mentioned at the beginning of this chapter, there are more than 200 papyri which throw light on problems of mechanical irrigation, and from this material I have selected 24 that I consider to be of value to our subject. For the information of readers, I include below a list of references to the papyri that have been excluded from this work. All of these have been published since 1920:

> *VII. P. Heid.IV No.95* (*P.Baden 95*) (1924); *VI–VII.SBGU 5300*; *VI.P.Vindob.Sijpesteijn 10* (1936); *VI.P.Oxy.2197* (1941); *VI.P.Gron.*

13 (1933); *VI.SBGU 9638* (1958); *589.P.Ross.Georg.III,40* (1930); *581.P.Oxy.1892* (1924); *569.P.Hamb.23* (1924); *555.P.Oxy.1913* (1924); *555.P.Oxy.2015* (1924); *528.P.Oxy.1900* (1924); *526.P.Michael.43* (1955); *V.SBGU 9137* (1952); *V.P.Herm.Rees 62* (1964); *V.P.Herm.Rees 38* (1964); *V.P.Ross Georg.55* (1930); *476.P.Oxy.1899* (1924); *449.SBGU 9515* (1959); *III.P.Oxy.1674* (1920); *III.PSI 808* (1925); *III.PSI 1072* (1929); *III.P.Michael.19* (1955); *III.SBGU 9473* (1961); *296.P.Mich.221* (1936); *255.SBGU 9410* (1959); *226.P.Oxy.2137* (1927); *II.P.Strasb.176* (1963); *II.P.Merton 79* (1959); *II.P.Mil.Vogl.69A* (1961); *II.P.Gron.19* (1933); *197–222.SBGU* 9541 (1958); *144.PSI 921* (1927); *141.P.Ross.Georg II,19* (1929); *II BC.BGU 1258* (1922); *II BC.P.Cornell 5* (1926); and *256 BC.P.Cair.Zen.59155* (1926).

However, the search must not end here. New papyri will appear each year, and among these there are bound to be a few that mention sāqiyas.

Vignette No. 7. Yoked oxen.

2. EXCAVATIONS

V–VI Abu Mina Egypt

(1) Ox-driven sāqiya working a potgarland on an elevated shaft.	Der große Doppelbad, G_1	p. 133
(2) Treadwheel and potgarland on common shaft.	Der große Doppelbad	p. 133
(3) Ox-driven sāqiya working a potgarland on an elevated shaft.	Der große Doppelbad, G_2	p. 134
(4) Ox-driven sāqiya working a potgarland on an elevated shaft.	Der Palast	p. 135

II–III Ostia, Italy

(5) Treadwheel with compartmented rim.	Terme del Foro	p. 137
(6) Two treadwheels with compartmented rims.	Insula dell'Invidioso	p. 137
(7) Two old treadwheels with compartmented rims.	Terme del Mithra	p. 137
(8) Later treadwheel with compartmented rim.	Terme del Mithra	p. 139
(9) Potgarland.	Terme della Trinacria	p. 140

I Tuna al Ǧabal, Egypt

(10) Ox-driven sāqiya working a potgarland on a long shaft.	Tuna al Ǧabal	p. 141

I B.C. Pompeii, Italy

(11) Treadwheel turning paternoster chain.	Terme Stabiane	p. 150
(12) Treadwheel turning paternoster chain.	Terme del Foro	p. 150
(13) Treadwheel turning paternoster chain.	Casa della Regina d'Inghilt.	p. 150
(14) Treadwheel turning paternoster chain.	Foro Triangulare	p. 150
(15) Treadwheel turning paternoster chain.	Reg. VIII, Ins. 5, No. 36	p. 150

I B.C. Alexandria, Egypt

(16) Tābūt with oxen.	Museum in Alexandria	p. 152

a. *Methods of investigating waterwheel ruins*

It is exceedingly difficult to find accounts of such works, and most of those I have found are limited to just a few lines. The best method seems to be to search for descriptions of wells. The archaeological literature contains thousands of these, but only those wells having a sufficiently large mouth for a potgarland are of interest.

Once the ruins have been located, the actual investigations can begin. The work is often impeded by shifting sands or grass, but as soon as this has been removed, the wells can be examined for signs of wear from the machinery. Such signs may take the form of broad, perpendicular and concave tracks in the lining of the well, corresponding to the pot diameter, or the shaft may have worn a deep groove in the brickwork, and the big wheel will often have left marks in the adjacent brickwork. Overshot waterwheels (cf. p. 154) and treadwheels (cf. p. 137) may have thrown such large quantities of water against the brickwork as to cause this to crumble, resulting in characteristic recessing of the end walls.

Fig. 90 and Plate 7. Sāqiya ruins at Abu Mina. Within this area, three different machines have worked at the same well at various times. See Figs 91a and 91b.

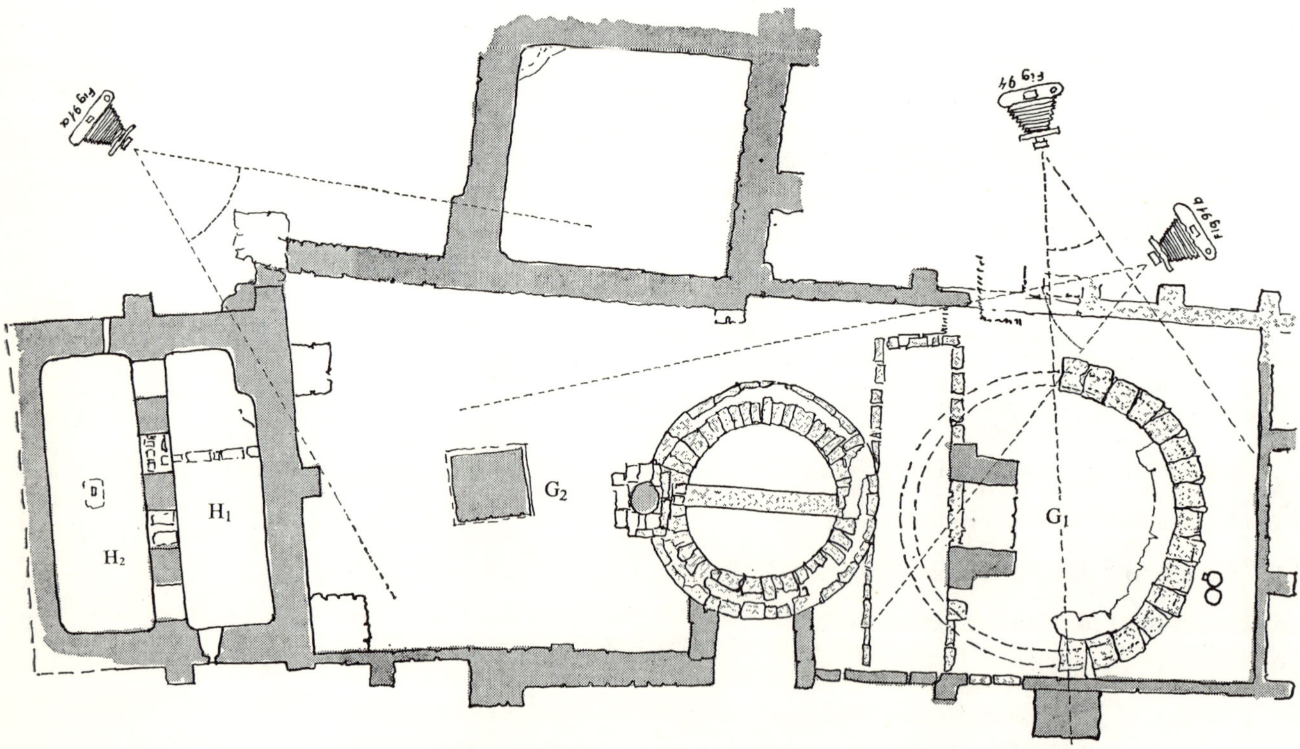

Lime encrustation provides us with another excellent clue, whereas the presence of green algae may simply indicate the original water level or the presence of stagnant water. At Tuna al Ğabal and Abu Mina, I found a heavy black coating on the wall, left by some tar product with which the woodwork had originally been treated.

Scale drawings are important, but before starting on these it is necessary to note any rebuilding that may have been done. My own experience has shown that this is an incredibly difficult task since all the works I have visited have been altered not once, but many times. If time is limited – and it usually is – a good camera will prove invaluable because it "sees" far more than the human eye in the strong sunlight.

b. *The sāqiyas at Abu Mina*

V–VI
Egypt

This ruined holy city is situated 75 kilometres west of Alexandria. It is reached by driving via Ikingi Maryūṭ to the little railway station at Baḥīğ. The ruins lie about 11 kilometres south of this, on the other side of the railway line.

At the turn of the century, two wells were excavated at these ruins, but it was not until Professor Wolfgang Müller-Wiener's publications that it was realized that they were wells for sāqiyas. W. Müller-Wiener: *Abu Mena.* [Mitteilungen des Deutschen Archäologischen Instituts, Abteilung Kairo]. An index map on p. 207 of Vol. 22 of this publication shows *Der große Doppelbad* and *Die Brunnenanlage Nord* or *Der Palast.*

In the same journal Vol. 20, facing p. 127, there is a plan of one of the

Fig. 91 a. Sāqiya ruins at Abu Mina. The plinth at G_2 is the centre of a circular ox-track. G_1 is the centre for another ox-track.

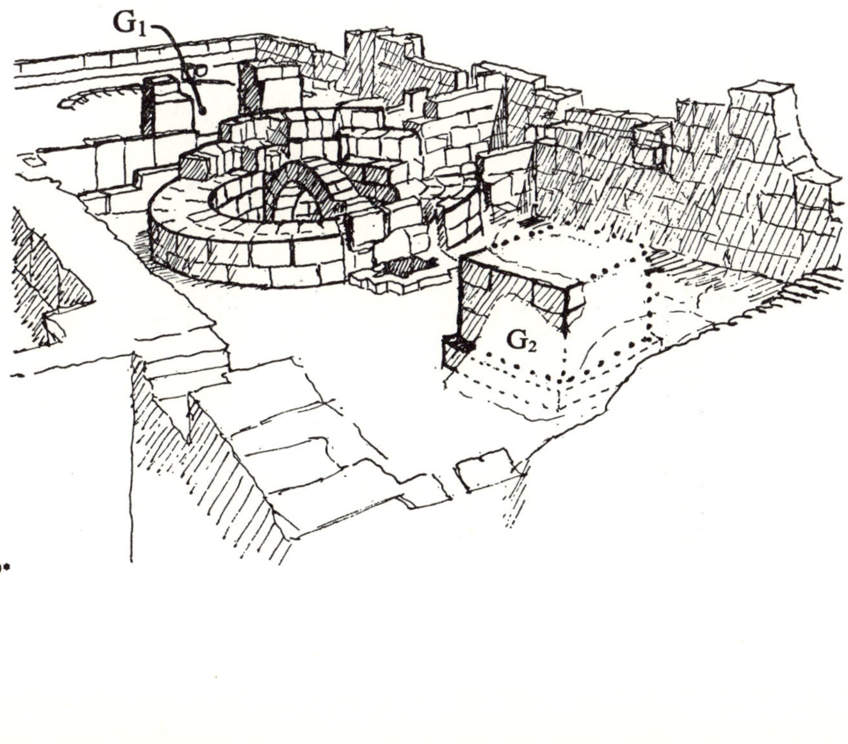

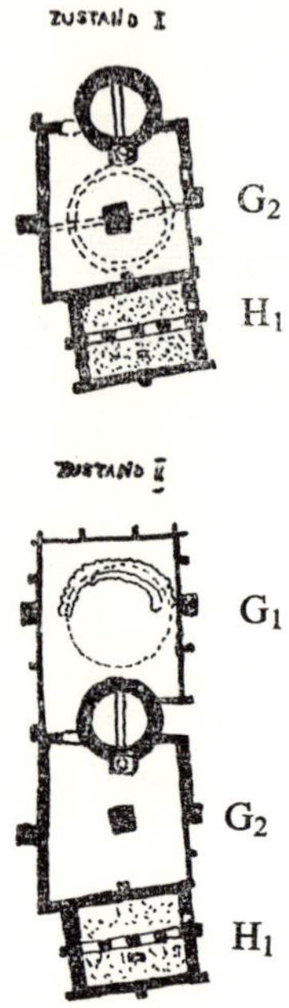

Fig. 91 b. Sāqiya ruins at Abu Mina. The hatched wall at the front of the picture is the wall of the trench. This served as pit for a treadwheel.

Fig. 92 and Plate 8. Abu Mina. Professor Wolfgang Müller-Wiener's reconstruction of the first and second machines, from the 5th and 6th centuries, respectively. Machine No. (3) and machine No. (1).

works, which is copied here as Fig. 90. The well is located between G_1 and G_2. A perspective of the works is shown in Figs. 91 a, 91 b. However, we are faced here with two periods, as can be seen from Vol. 21, p. 176 (Fig. 92 is a copy). On the basis of these drawings and a cross-section drawing – most kindly placed at my disposal by Professor Müller-

Fig. 93. Abu Mina. Reconstruction of a sāqiya with elevated shaft. Machine No. (1). (cf. Fig. 92, Zustand II).

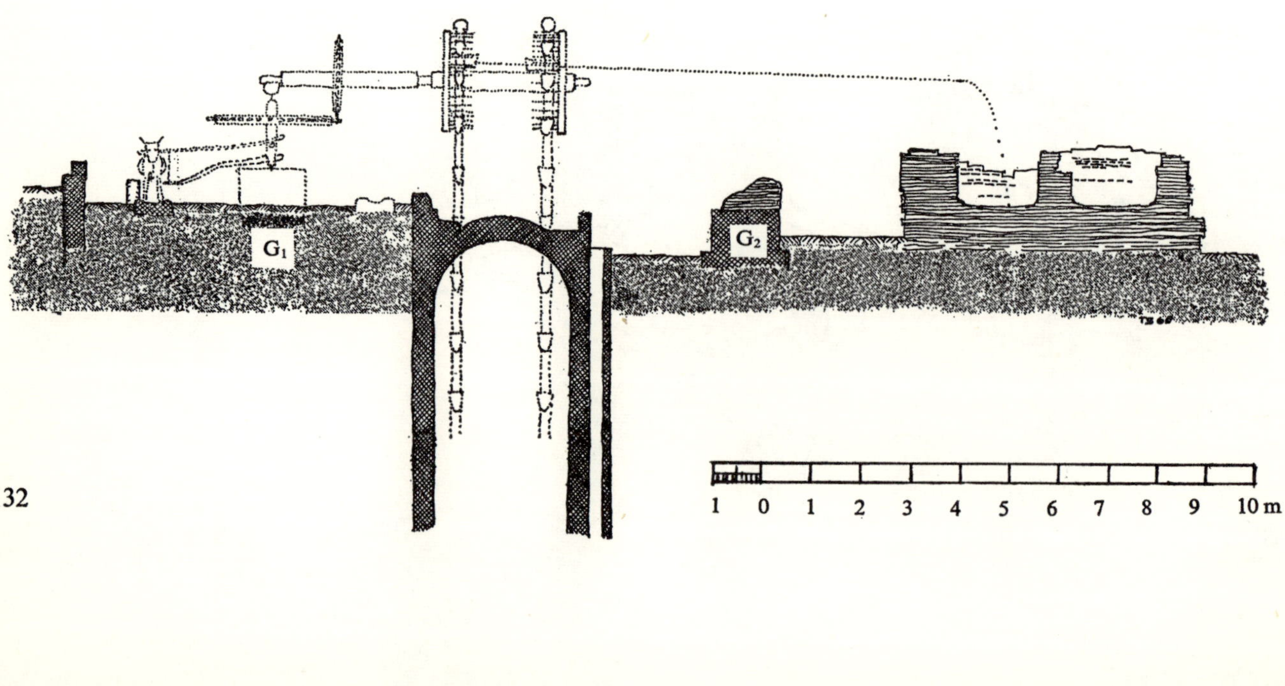

Fig. 94. Abu Mina. The circular ox-track. The cylindrical stone must have been a bollard to which the drawbar was fastened when the ox was to be unharnessed.

Machine No. (1)

Wiener – I have reconstructed two of the four machines that once operated at this sanctuary. The reconstruction, which is shown in Fig. 93, is based on the circular ox-track, Fig. 94 at G_1, and four perpendicular grooves in the ashlar lining of the well caused by the hoisting movement of the potgarland. I have also assumed that the two cistern walls at H_1 and H_2 were originally higher than they are now.

For some unknown reason, the system was at some time changed from ox-driven machinery to treadwheel machinery. Half of the circular track was removed, and a trench was built right beside the well. On the other side of the trench, two pillars were raised.

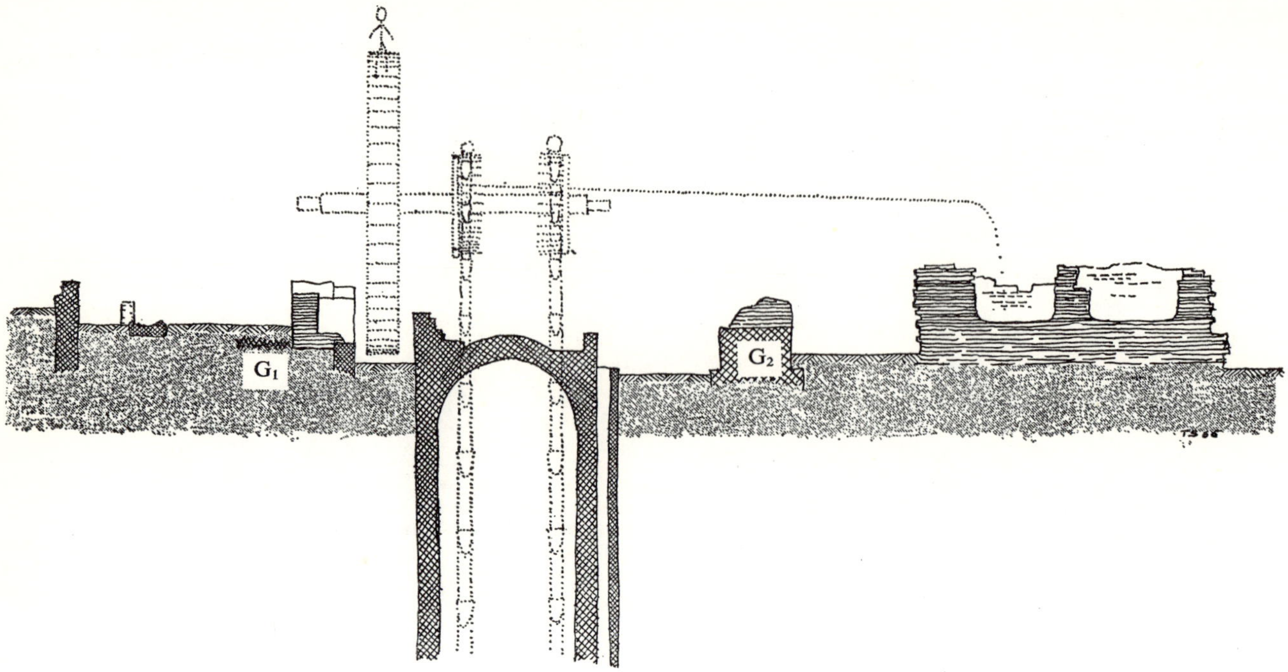

Fig. 95. Abu Mina. One of the machines that once operated here was a paternoster chain turned by a treadwheel. Machine No. (2).

Machine No. (2)

For a reconstruction of these works, see Fig. 95.

During a visit to Abu Mina in October 1964, I familiarized myself with the trench beside the well, but its presence there remained a mystery to both Professor Müller-Wiener and myself. After I had returned home it occurred to me that it might be a pit for a treadwheel. I informed Professor Müller-Wiener of my supposition, and a fortnight after the commencement of his new season (1965), he wrote to me as follows:

> ... Bei näherem Ansehen ist es aber klar, das da etwas neugebaut worden ist – also das von Ihnen vorgeschlagene Tretrad.

Four days after the conclusion of the excavation season I received further information:

> ... aber einleuchtend ist der Vorschlag mit dem Tretrad schon. In der Grabung habe ich übrigens noch die Fundamentierung des großen (Zahn)rades untersucht: In der Mitte der "manège" liegt ca. 1 m unter dem Boden ein dickes Mauerfundament, das z.T. abgestürzt zu sein scheint bzw. abgerissen und das man später einfach zugeschüttet hat – d.h. man brauchte dieses Mittel-Lager nicht mehr – ein Beweis mehr für den Vorschlag des Tretrades.

Machine No. (3)

As will be seen from Professor Müller-Wiener's letter, the remains of a foundation were found under G_1, Fig. 90. On the other side of the well, just by G_2, there is another, partly excavated plinth. In 1964 its size convinced me that this was not part of a third sāqiya, but since my visit to the Near East I have changed my mind; this must be the base for the bedding stone for a sāqiya.

At several museums there I saw "door socket" stones, some of them measuring as much as 36 × 36 centimetres and with a height of 22 centimetres, and others measuring only 8 centimetres in diameter, but all with a cup-shaped socket. Further examples of such stones are included in (*Fondation Carlsberg*. Hama, Fouilles et recherches 1931–1938 II_1) E. Fugmann: *L'Architecture des périodes pré-hellénistiques*. Copenhague 1958, fig. 137, p. 110. I know that these stones were used as door gudgeon sockets, but they had other uses as well; for instance, 10 metres from the water mill in the Athenian agora, (cf. p. 154) there is an oil mill, at the centre of which there is a large, circular "door socket" stone cast in position by means of lead. However, most "door socket" stones have nothing at all to do with doors but are actually the bedstones of small grinding mills. Max Freiherr von Oppenheim: *Tell Halaf*, Vol. 4. Berlin 1962, p. 51. Plate 39 c illustrates both the bedstone and the runner of a pivot-grinding mill (Zapfenmühle). The question is whether such stones were also used for sāqiyas; the answer may well be waiting for us under the layers of sand and stone covering the plinth at G_2.

Machine No. (4)

The last sāqiya-works are located at *Der Palast*, see Fig. 96, which is a copy of the drawing in Vol. 22, facing p. 208 and [Archäologischer Anzeiger 1967, facing p. 459.]

The works here consists of a 21 metre deep well, the ashlar lining of which has four grooves in it, worn by the hoisting movement of the potgarland. In the middle of the adjacent yard there is a plinth made of a mix-

Fig. 96 and Plate 9. Abu Mina. The sāqiya-works towards the north. The remains of the plinth for the bedding stone lie between the well and the water tank. Machine No. (4).

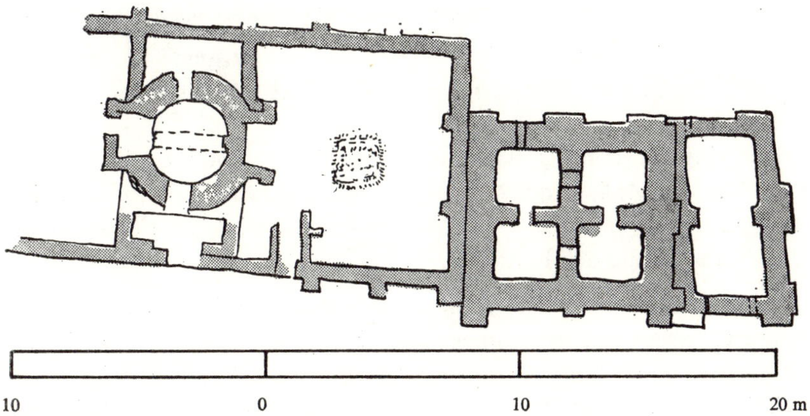

ture of lime and pebbles. Although its shape is somewhat irregular, I believe that this was once the base for the bedding stone for a sāqiya.

There is a picture of this well in Karl [Carl] Maria Kaufmann: *Die Menasstadt und das Nationalheiligtum* .... Leipzig 1910, plate 50. In this a number of pots can be seen lying on the edge of the well, some of which resemble qādus-pots.

Fig. 97. Terme del Mithra at Ostia, Italy. The wall showing deep grooves worn by a water-lifting wheel.

Apart from four rather puzzling works at Pompeii, this is possibly the only place at which there is a treadwheel working a paternoster chain. A detailed account of such works has been left by Vitruvius, Book X, Chapter V, Section 4. A. G. Drachmann: *The Mechanical Technology of Greek and Roman Antiquity*. Copenhagen 1963, p. 151:

> But if it is to be delivered in higher places still, round the axle of the wheel there is placed a double iron chain let down to the lowest level, with hanging bronze buckets on it, each holding a *congius* (3.3 litres). So the turning of the wheel, by winding the chain round the axle, lifts the buckets up on top, which, when they are borne up over the axle, are forced to turn over and pour out into a container as much water as they have lifted out.

c. *Waterwheels at Ostia*

As mentioned in the above section, the sāqiya at Abu Mina had been replaced by a treadwheel, although the paternoster chain itself was retained. Here, we shall examine a different type of treadwheel, namely, that with a compartmented rim. Such wheels were used in Ostia, and the best description we have of them is that by Russell Meiggs: *Roman Ostia*. Oxford: Clarendon Press 1960, p. 144:

II–III A.D.
Italy

> To supplement supplies in the baths water-wheels were installed to fill the cisterns from the subsoil[2]; in many private houses wells were still retained.
>
> Note (2): The clearest traces of such water-wheels can be seen in the Baths of Mithras and in the Forum Baths; few sets of Ostian baths are without them.

Report

In July 1965 I visited Ostia in the course of a tour made possible by a grant from *Statens almindelige Videnskabsfond* (Ministry of Cultural Affairs).

Following the main street of Ostia, the Decumano Massimo, as far as the Via dei Molini and then turning south down the Via dei Semita, I arrived at two rectangular wells, one at the Terme del Foro and the other in the Insula dell'Invidioso.

Machine
No. (5)

At the first site I had great difficulty in forming an idea of the original structure, except that the dimensions of the well indicated that it had once contained a wheel of about 6 metres in diameter. The series, *Scavi di Ostia*, Vol. *I*, Roma 1953, contains a unique set of plans of Ostia, and on No. 8 it is just possible to distinguish this well and its water tank. As will be seen in the next example, such water tanks may show traces of wear from the treadwheel, but I found none at the first site.

Machine No. (6)

My inspection of the second site, the baths in the Via dell'Invidioso, Regio V, Insula 5, was far more fruitful. In one corner of a roofed-in room measuring 6 × 3 metres there is the characteristic recessing caused by the spray from the waterwheel washing the mortar out of the wall, whereby the stones have fallen out. The cavity in question – there are several others in Ostia – is 50 centimetres wide and several metres in height. See Fig. 97. From this room, which has only been partially excavated, the water was raised to a water tank measuring 3.85 × 0.8 metres and 4 metres in height. The bottom of the tank lies 2.7 metres above the water level in the room, which means that the first wheel must have had a diameter of somewhat more than 5 metres.

In the mortar layer of the water tank there are clear traces of yet another wheel. These traces have a radius of 1.8 metres, which means that the diameter must have been about 3.6 metres. The centre of this circle is located about 2.1 metres above the bottom of the water tank, so the two wheels must have had a total lift of about 7 metres.

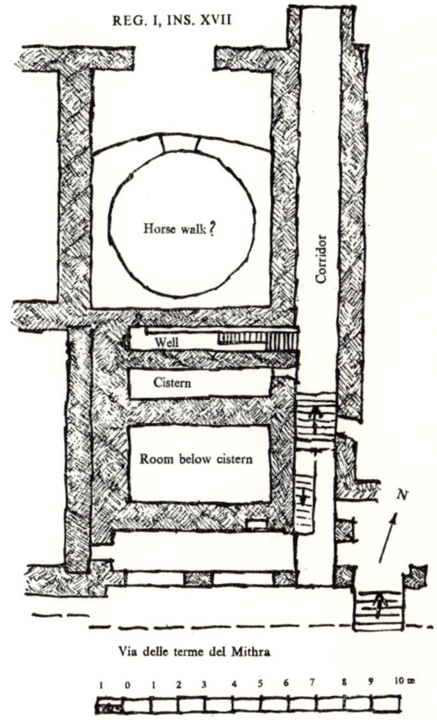

Fig. 98. Plan of the hydraulic works at Terme del Mithra.

Our visit to Ostia does not end here. If we now return to the Decumano Massimo and follow this road to where it forks and there take the right turning, we come to the Via delle Terme del Mithra, where we find the best preserved water-lifting system in Ostia.

My work here was greatly facilitated by the fact that Professor Th. Kraus of the *Deutsches Archaeologisches Institut*, in Rome, had recommended me to the *Soprintendenza alle antichità Roma no. III*, which immediately issued me with a permit to measure and inspect the Terme del Mithra.

The hydraulic system here is extremely complicated because the plant has been altered several times. I have therefore returned to the site twice since 1968, but there are still a number of details that need clarification.

Entering the baths from the Via Terme del Mithra, we find ourselves in a long corridor, which leads to the parallel street on the other side. At right-angles to the corridor there are two long chambers, the first of which is a water tank, and the second, a well. See Fig. 98.

Machine No. (7)

In the mortar layer of the well there are marks worn by big waterwheels dating from two different periods. The oldest wheel (Figs. 99 and 100), which had a diameter of 7.5 metres, raised the groundwater to a water tank, the position of which is indicated by the number 14 in Fig. 99. However, there is a staircase in the well, the top six steps of which go from wall to wall. This seemed to preclude the possibility of the oldest wheel ever having been located here, but to make quite sure, I went under the stairway, and there I not only found clear traces of

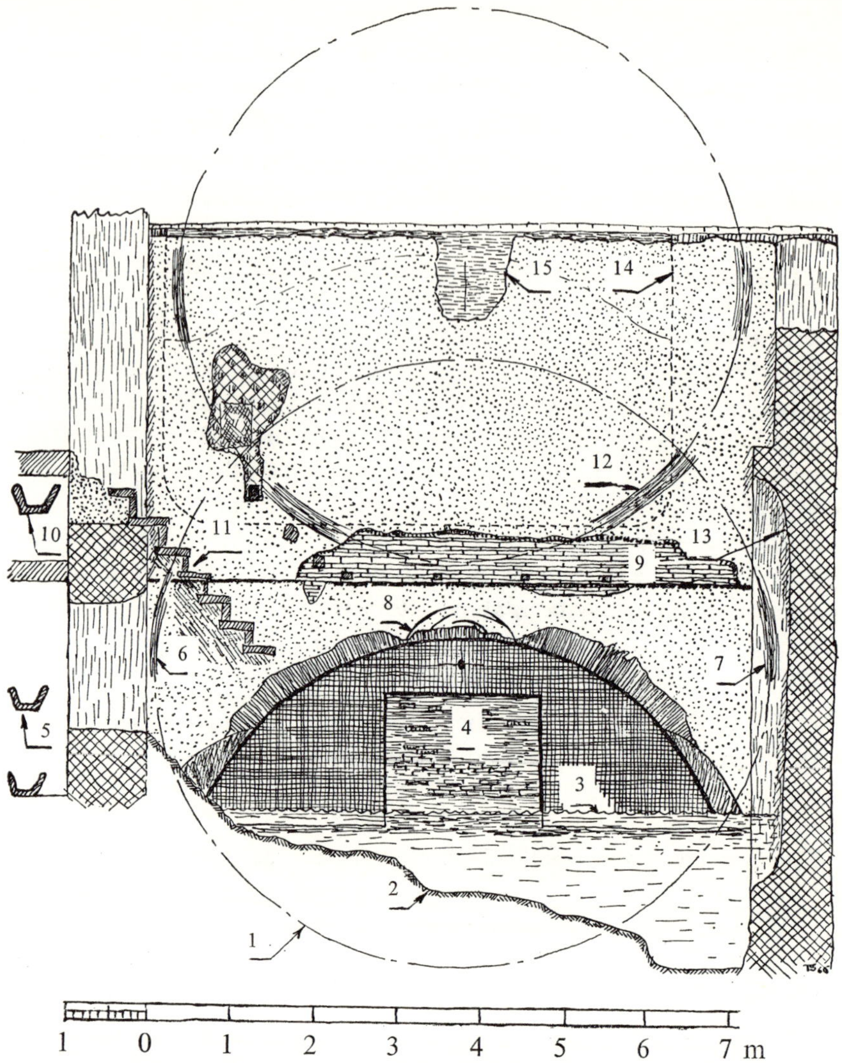

Fig. 99. Terme del Mithra. Grooves left by water-lifting wheels. See texts to labels on facing page.

wear, see Fig. 99, No. 6, but also discovered that the six steps in question were of recent date. The same wheel has also made some very deep grooves near the centre of the wheel.

From the width of the well it appears that the wheel cannot have had a width of more than 45 centimetres, which in turn indicates that it must have been worked by men treading on top of it; there can certainly not have been enough space for them to walk inside the wheel, and even supposing there were, they would have been constantly drenched with water. Similar top-tread wheels were used along the river Mondego in Portugal right up to the present century. Jorge Dias: *Aparelhos de elevar a água de rega.* Porto 1953, pp. 154 and 177.

The next point to be clarified is the height to which the water was raised. There can be no doubt that it was lifted to the water tank (No. 14) located behind the wall, but I did not manage to find the inlet. This may be explained by the fact that the wall of the tank shows signs of later bricking-up.

I searched the tank in vain for signs of a second waterwheel, similar to that in the Via dell'Invidioso. Unfortunately, it lacked the hard mortar layer in which there might have been wear marks corresponding to those on the other side of the wall.

Machine No. (8)

Now we come to a new era in the history of Ostia, heralded by a great innovation – the canalization of water direct from the street to the houses. The old wheel was removed, the stairway mentioned above was built, and a new wheel, 6.7 to 6.8 metres in diameter, was installed 5 metres higher than the original wheel. The later wheel has left very clear marks in the hard mortar layer. See Fig. 99, No. 12. On both sides of the well we can still see the holes left by the four transverse beams that once carried the trough from which the wheel lifted the water to a tank on the present roof of an adjacent building. I did not succeed in actually proving that there had been such a water tank up there, but, from the solidity of the vaulted roof, it is obvious that this could well have carried a water load of 15 to 20 tons.

The east wall of the well in the Terme del Mithra, surveyed by the author in 1965. See Fig. 99.

1 Oldest wheel
2 Bottom of the well, 1965
3 Water level, 1965
4 Wall or pier for bearing. See Fig. 97
5 Gutter, probably for return-water from the baths
6 Marks under new staircase
7 Marks in niche
8 Marks from old wheel
9 At this level a new trough or basin was built. The water was supplied by the gutter, No. 10
10 Water gutter from street
11 New stairs for the staircase
12 Clear marks from the new wheel
13 Niche or recessing. See Fig. 97
14 Line of cistern behind the wall
15 Location of bearing for the axle

As will be seen in Fig. 98, there is a large, circular platform on the other side of the well, which it seems reasonable to assume is the remains of a horse-walk; the platform has a diameter of about 5 metres, which suggests

Fig. 100. Terme del Mithra. Reconstruction of the first hydraulic works. Machine No. (7).

this possibility. Unfortunately, the whole thing was so overgrown with high grass and brambles that I was unable to examine it properly, but I hope to have the opportunity of doing so on my next visit. From a technical point of view, the structure could well have been the platform of a horse-walk, but I have not seen the like elsewhere, and hesitate to make the suggestion. [I visited Ostia again in 1971, and this time the platform had been cleared. It proved to have nothing at all to do with a horse-walk, but seemed instead to have been some kind of chapel].

Machine No. (9)

Not far from the Terme del Mithra we find the Via del Serapide, on which there is yet another baths, the Terme della Trinacria. Here, behind a modern railing there is a shallow, rectangular well measuring 5.6 × 1 metres and 2.7 metres in depth. However, there are several indications that the present groundwater level is 1 metre higher than the Roman water table.

In this well there are two deep grooves that have obviously been caused by the constant scraping of a potgarland against the wall. See

Fig. 101. Both grooves, which evidently differ in age, must have been caused by a very heavy, ascending potgarland, the pots of which had a diameter of about 20 centimetres.

There is, of course, the possibility that the grooves might have been caused by a bucket on a well-rope pulled up over a pulley-wheel, but such a bucket hanging free at the end of a rope would swing from side to side, thereby wearing fan-shaped marks in the wall. A heavy potgarland swings very little because it is constantly in water, which has a very heavy, damping effect. A potgarland hangs straight down, even when it is in operation. The four potgarland grooves found in the well at Abu Mina had the same depth and shape as those found here. The unpublished volume of *Scavi di Ostia* will probably provide us with more information on the waterwheels of Ostia. I. Gismondi: *Gli impianti idaulici.* Roma 19 .....

Fig. 101. Ostia. Wall with deep grooves worn by a potgarland. Machine No. (9).

d. *Hydraulic works at Tuna al Ǧabal*

These works, which were built at about the time of the birth of Christ, are among the finds that best elucidate the earliest annals of the sāqiya, and I am deeply indebted to Professor Wolfgang Müller-Wiener for drawing my attention to them.

To get to the site, I took the morning train from Cairo to Māllawī, about 300 kilometres to the south. From there, a taxi took me the rest of the way, past the ruins of the ancient Hermopolis, through innumerable sugar plantations, and out into the desert, where, after driving past a number of rocky outcrops, I finally reached Tuna al Ǧabal.

Here, in 1936, under the direction of Professor Sāmi Gabra, an enormous well system was excavated, built entirely of burnt red bricks, while the temple yard in which it is situated is of white sandstone.

Machine No. (10)

The first description of this excavation was given by Professor Sāmi Gabra: [The Illustrated London News, June 12, 1937, No. 5121, Vol. 190, pp. 1088–1089], and here we find the following information:

The works are from about A.D. 100, but they became overlaid with sand as early as the 3rd century. They were originally surrounded by a fertile garden sheltered by palm trees, and the trunks of these can still be seen. The description is accompanied by a number of photographs giving an impression of the site. (I myself have several photos of the waterworks).

Unfortunately, part of the works had been restored and a new sāqiya built, in honour of His Majesty, King Fārūq, who visited the excavations and saw the plant in operation on the 7th of January, 1937.

In another paper by Professor Gabra: *Fouilles de l'Université "Fouad el awal" à Touna el Gebel* (*Hermopolis ouest*), there are fifteen lines describing these hydraulic works. [Annales du Service des antiquités de l'Égypte, Vol. 39, p. 489 ff. and plate *LXXXIII* (a photograph of the reconstructed works)]. The same lines appear again in Bulletin de la Société française d'égyptologie, No. 30, 1959, p. 49:

> .... Ce jardin est arrosé .... This garden is irrigated by a big waterwheel (a saqieh) operating in two stages and excellently preserved – indeed, a masterpiece of architecture. The two wells have a total depth of 34 metres. The problem of lifting the water from such a depth is solved in an ingenious manner. In the lower well, the water is lifted by hand and poured into a basin leading to a second basin, which is rectangular in shape and 14 metres deep. From here the water is lifted by a potgarland and a wheel powered by an animal. A spiral staircase leads down to the first platform of this big well.

The paper also treats the other finds made in this huge excavation area. The hydraulic works themselves are only a tiny detail of the necropolis of Hermopolis.

In 1956, the works were surveyed and described by Alexandre Badawy, who is an architect: *Au grand temple d'Hermoupolis-Ouest: l'installation hydraulique.* [Revue archéologique, 6e série, Vol. 48 (Paris 1956), pp. 140–154]. This paper, which includes eight magnificient drawings, contains many fine details of the well and the stairway, and describes similar spiral staircases in other Egyptian architecture.

The best way of describing this imposing structure is to follow the course of the water from the bottom of the well to its final discharge. However, as the design is rather complicated, the reader would be advised first to study Fig. 102.

As mentioned above, the water was lifted in two stages, the first of which comprises a 3 metre diameter, 20 metre deep well. In the lining of the well there are a number of recesses for transverse beams, but there are no signs of wear by either paternoster chains or machinery. Around the lip of the well there is a circular walk containing two 20 centimetre deep basins, one of which has an outlet leading to the bottom of a 14 metre high rectangular well shaft. However, before describing this in detail, we must take a look at two vaulted recesses in the wall, 2 metres up from this circular walk. These recesses serve as supports for a beam, but they have unfortunately been filled in with new bricks and mortar, which prevented me from examining them in detail.

The beam, which dates from 1936, is provided with two well windlasses. These are not shown in my drawings but can be seen in the paper by Badawy. I must confess that I at first doubted the authenticity of this reconstruction, but I have been unable to find a better solution, and today I am inclined to agree with that of the local sāqiya-carpenter. This type of windlass is common in Egypt (see Fig. 60) and was also known in antiquity; there are two at Herculanum, one of them in its correct position, and they are frequently depicted in Coptic art.

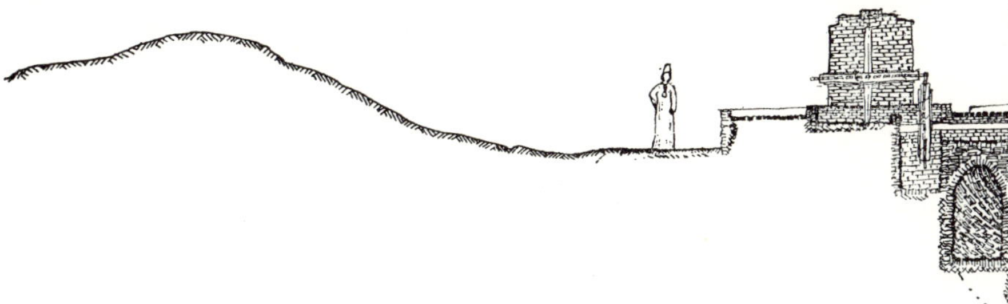

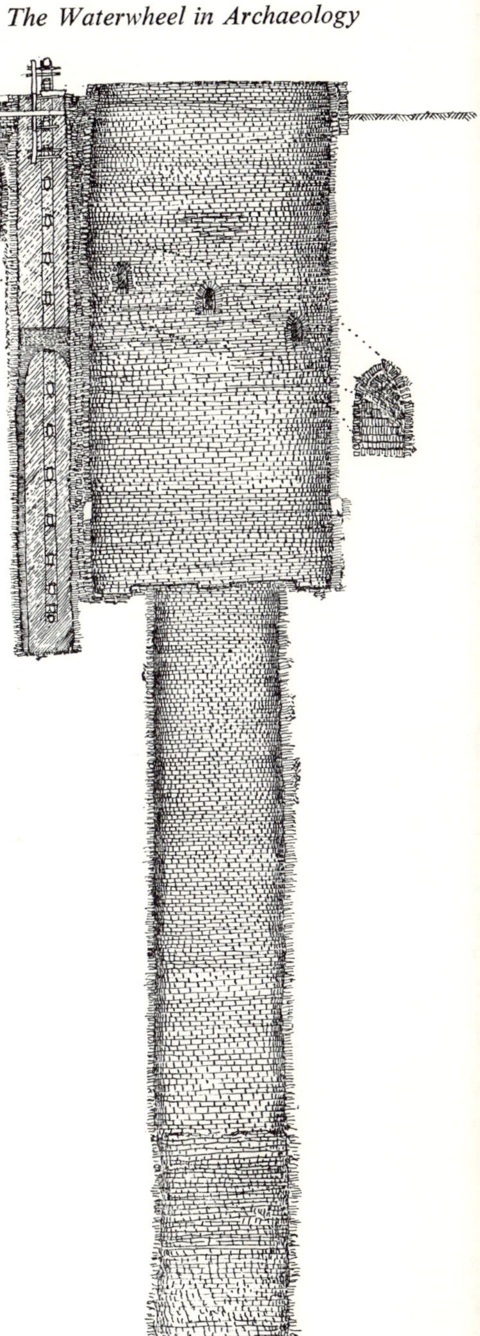

Fig. 102. The Roman hydraulic works at Tuna al Ǧabal in Egypt. The water was raised from a 20-metre deep well and emptied out into two small basins, from which it was carried through a number of narrow pipes to a water tank. A sāqiya then raised the water up the last 14 metres to a trough, from which it flows to another water tank and then to a couple of taps located under a domed arch. The horizontal shaft of the sāqiya has left deep grooves in the shaft tunnel. The works were partly restored in 1936. From the first century B.C.

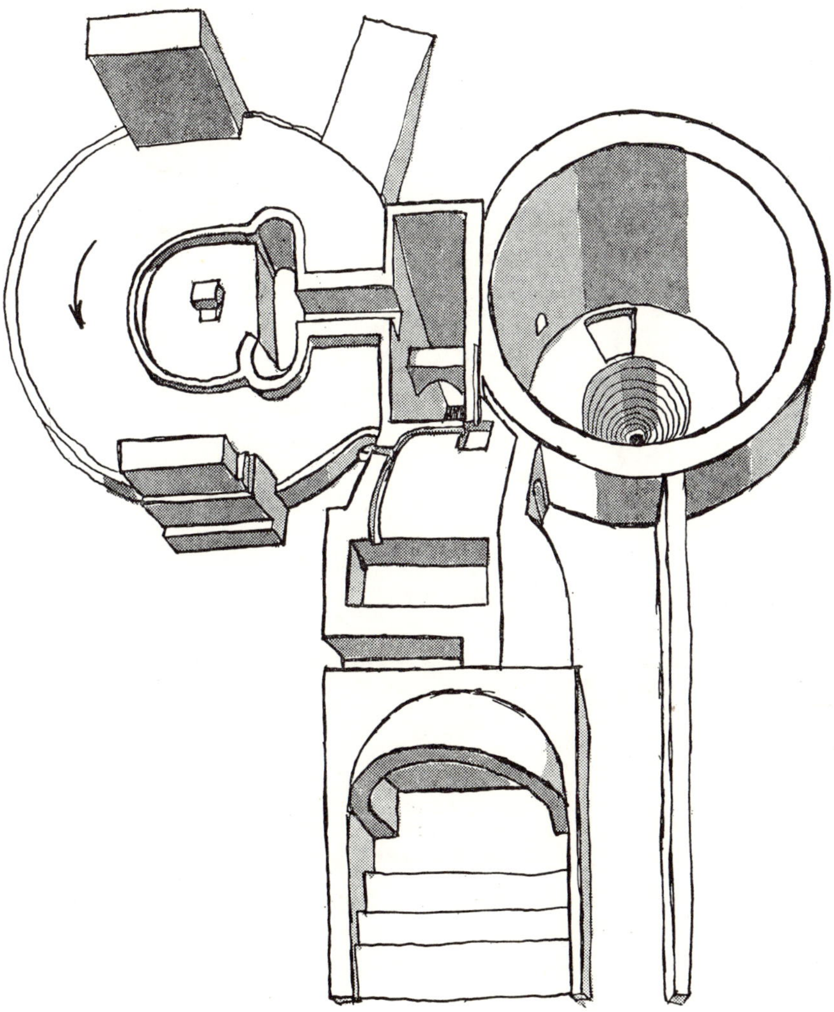

Fig. 103. The reconstructed sāqiya at Tuna al Ǧabal. Machine No. (10).

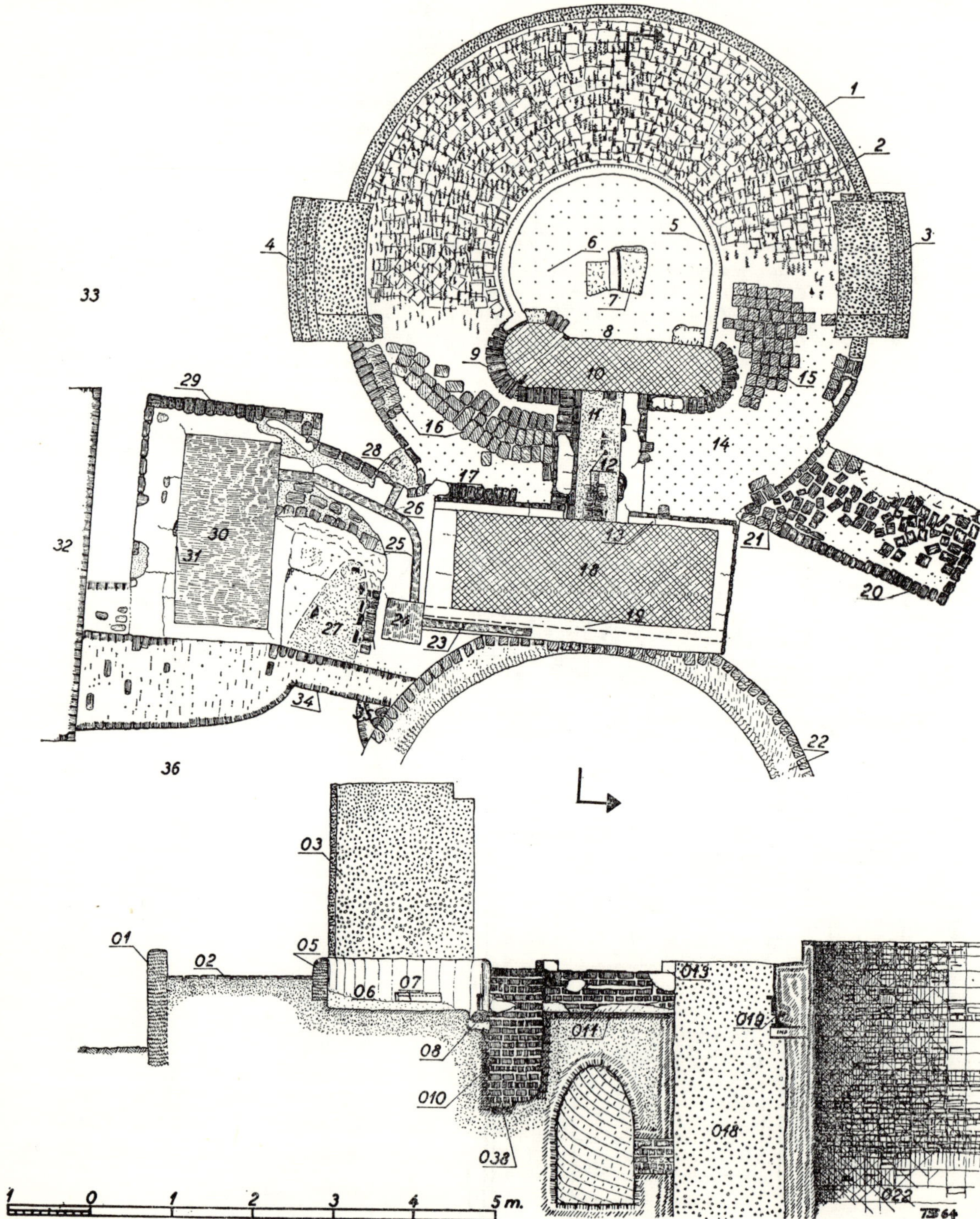
1
2
3
4
5
6
7
8
9
10
11
12
13
14
15
16
17
18
19
20
21
22
23
24
25
26
27
28
29
30
31
32
33
34
35
36
03
01
02
05
06
07
08
010
011
013
018
019
022
037
038
1 0 1 2 3 4 5 m.
TB 64

Plan of the Roman waterworks, surveyed by the author in 1964.

1 Wall of mud brick and daub
2 Circular track paved with mud bricks and drift sand
3 Pillars of red brick, plastered with new mud. Arabic inscriptions
4 Both pillars are new; the base may be old
5 Circular wall of mud brick and daub. Of recent date
6 Covered with drift sand. Not examined
7 Sandstone bedding for upright shaft. Probably the old bearing for the old horizontal shaft
8 Modern bearing of iron and wood, fixed in position with gypsum
9 Old bricks, but recent construction. Modern sandy mortar
10 Cog-wheel pit. The wall between Nos. 10 and 11 is old except for the upper part. A large part of the wall was cut away in 1936 to make room for the reconstructed wheel. This wall is not connected to the adjoining staircase wall. See arrows
11 Shaft tunnel. Red brick, blackened by pitch. Top courses of recent date
12 The old shaft has worn a deep groove in the brickwork. See Fig. 105
13 Modern well-lip. Plastered with grey cement
14 No remains of mud bricks, only sand
15 Mud bricks of surprisingly soft consistency
16 Mud bricks as found in 1963. Most of them disappeared during 1964
17 Very old brickwork
18 About 12 metre deep shaft with cistern at bottom
19 A modern steel beam supports the shaft-lip
20 Old gangway, partly destroyed
21 Mud bricks
22 The big well-wall. Burnt brick from top to bottom
23 Gutter coated with cement
24 Small tank, plastered with grey cement
25 Gutter, about 6 centimetres deep
26 Overflow gutter for reservoir
27 Top of arch belonging to spiral staircase. No. 28 *idem*
29 Burnt red brick with deep wear marks. Photo available
30 Reservoir, two coats of white and grey cement plaster
31 Inlet for pipe
32 Domed fountain. Brickwork in herringbone pattern
33 Remains of palm-trunk.
34 Modern door to spiral staircase
35 Worn red brick steps
36 Path down to door
037 Spiral staircase with peep-hole in shaft
038 Mud bricks coated with plaster. Part of the bricks cut away

← Fig. 104. Survey of the Roman sāqiya ruins at Tuna al Ǧabal.

Although there is some doubt as to how the water was lifted the first 20 metres, it is certain that a sāqiya was used for the last 14 metres. The potgarland hung in a rectangular shaft, the bottom of which formed a water tank measuring 3 × 1.45 and 1.7 metres in depth. At the middle of the shaft there is a supporting arch for the walls. See Fig. 103. On one side of the shaft, just above the water tank, there is a door opening, from which it is possible to jump down into the tank, but there are unfortunately no signs of potgarland wear here either.

If we now leave the bottom of the shaft and walk up the spiral stairway, with the well on our left, we reach the top of the well, from which we can follow the flow of the water. Here there is a small channel leading to a cistern from which the water passes through a pipe to four stopcocks located underneath a domed arch of unique archaeological interest.

Report

Investigation of the brickwork surrounding the Roman sāqiya

I am indebted to *Statens almindelige Videnskabsfond* (The Ministry of Cultural Affairs) for a grant enabling me to undertake this investigation and to prepare a scale drawing of the works in October 1964.

The ruins around the sāqiya are extremely well preserved, but part of the works must have been reconstructed. It is very difficult to determine how much of the brickwork dates from Roman times and how much from 1936. The red bricks are all from the former period, but in some instances they have been built up again with new mortar, while in others, the brickwork is old but has been re-pointed with new mortar. In several places, the brickwork is covered by a new layer of mud daub; in some cases, plaster has been used, while in others, there is still the original white Roman cement. However, wherever the mortar is white and porous, we can be sure that the brickwork is Roman.

The scale drawing is reproduced in Fig. 104.

After removing the drift sand from the cogwheel pit (see frontispiece), I examined this carefully, expecting to find marks worn in the brickwork by the cog-wheel. My hopes were frustrated when I found that a 20–30 centimetre layer of brickwork (between No. 10 and 11) had been chiselled away. This wanton destruction must have been the handiwork of the local sāqiya-carpenter when he placed his modern sāqiya on the old machine foundation. In his reconstruction, the horizontal cog-wheel has a diameter of 3.2 metres, whereas that of the original wheel can hardly have exceeded 2.4 metres. Therefore, in order to make room for the new geared structure, the carpenter first placed the centre of the wheel 20 centimetres off centre of the circular track and then cut away about 30 centimetres of the brickwork near the bottom of the pit.

Evidence of the fact that there really was once a sāqiya in operation here is found at the bottom of the shaft tunnel, Nos. 11, 011 and 12, where there is a deep, concave groove, extending over a length of more than 70 centimetres, ground into the brickwork by the original Roman shaft. In this groove there is a number of parallel scratches, which have been caused by small stones being ground round between the shaft and the brickwork. See Fig. 105. Such marks might perhaps be thought to have been caused by the new shaft from 1936, but this is fortunately still supported in its bearings high above the bottom and can therefore not have been responsible for these very rare traces.

The bricks at the bottom were originally red, but are now blackened by tar. The timberwork had probably been treated with tar, as described by Vitruvius, Book 10, Chapter 4, Section 2, and P. Lond. 1177 line 230.

Fig. 105. Tuna al Ğabal. Wear marks left by the horizontal shaft at the bottom of the shaft-tunnel. See Fig. 104, No. 12.

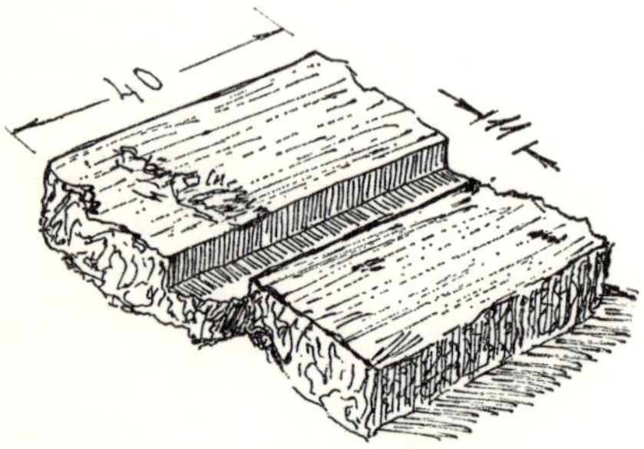

Fig. 106. Tuna al Ǧabal. Sandstone slab with groove for wooden bearing.

The horizontal shaft had two bearings but there is now no trace of these. On the other hand, within the kerb-stone circle, No. 5, I found a flat sandstone slab lying hidden under a layer of drift sand. This stone has a deep groove in it, in which a wooden block must once have rested. See Fig. 106. In such a case, the block will have served as a bearing for the horizontal shaft. However, as the stone was not found *in situ*, I must admit that the premises for this assumption are not as strong as they might be.

On the basis of the horizontal shaft and the rectangular well-shaft, it is possible to form an idea of the original structure. The paternoster wheel must have had an internal diameter of about 2.4 metres, and the small cog-wheel a diameter of about 2 metres. The big cog-wheel, on the other hand, cannot have been as big as the present-day sāqiya-wheel; I think it was probably no more than 2.4 metres in diameter. The circular ox-track must have had a diameter of less than 5 metres, which is also somewhat smaller than in the modern sāqiya.

There are indications that the flow must have been rather modest. For example, the outlets in the rectangular shaft are only 15–20 millimetres in diameter, and the small water tank, No. 30, has a capacity of 2 cubic metres. This is far below the figures from Pompeii, 38 cubic metres, and Quṣayr al ʿAmra, 18 cubic metres.

However, it is really incorrect to try to compare these figures, because we do not know how often the tank was filled up. The tank in Pompeii had to hold sufficient water for a large baths consisting of two big pools. At Quṣayr al ʿAmra, there was a small baths without any pools, and at Tuna al Ǧabal, the water was used for drinking purposes and for watering a few flower-beds, but in no respect an irrigation scheme.

There were fragments of unglazed pottery scattered about the site, and about 20 metres from the well lay a pile of broken pots. Several hundred pieces have been examined, but none of them seems to have come from sāqiya-pots. I myself am of the opinion that leather sacks were used instead of pots.

As it was of great importance to date these works, I removed about 10 grams of the stump of the palm tree located at mark No. 33 in Fig. 104. The sample was sent to The Copenhagen Radio-Carbon Dating Laboratory, which kindly carried out the necessary test. The test, which dated the sample to 1900 $\pm$ 100 years before A.D. 1950, is published in Henrik Tauber: *Copenhagen Radiocarbon dates IX.* [Radiocarbon, Vol. 10, No. 2, (1968), p. 321, under K-1283 and not K-1282.]

This dating means that the palm tree is from about the year A.D. 50, or, expressed in more precise terms, the probability that the tree died in the period B.C. 50 to A.D. 150 is 68 per cent. It can therefore be concluded that these works were built at the beginning of that period.

e. *Water-lifting machines at Pompeii*

I B.C.
Italy

At Ostia, the groundwater level is only a couple of metres below ground level, so a treadwheel sufficed to bring the water to the surface. In cases where the water had to be raised to a greater height, the problem was solved by simply arranging one wheel on top of another. However, it seems that the Roman engineers of Ostia were acquainted with the paternoster wheel, although we have no idea of the appearance of these machines.

Conditions at Pompeii are totally different. At the centre of the city, the groundwater lies about 25–30 metres below ground level, and the Pompeian engineers were therefore forced to use bucket chains when large quantities of water were needed. The water for the Roman baths was usually channeled through aqueducts, but where none existed, or when they had been destroyed by earthquakes, it was necessary to utilize the bucket chain powered by some kind of machinery.

The water-lifting machines of Pompeii have been treated by Rudolf Pemp: *Drei Wasserhebewerke Pompejis.* (Technical Dissertation 30–*XI*–1938, Dresden). Konrad Triltsch, Würzburg-Aumühle 1939. This is apparently available at only a few libraries, but both *Danmarks Tekniske Bibliotek* and *Bibliothek des Deutschen Archaeologischen Insituts*, Rome, possess a copy.

Pemp has investigated three wells at Pompeii in which potgarlands were once used, one at the Terme Stabiane, another at the Casa della Regina d'Inghilterra, and a third at the Terme del Foro. These wells are described in brief in Amedeo Maiuri: *Pozzi e condotture d'acqua nell'antica città.* [Notizie degli Scavi di antichità, Series 6, Vol. *VII* (Roma 1931), pp. 552–553].

The Terme Stabiane has a 20-metre deep well, in which a potgarland was once operated by a big wheel turned by trotting men, so no trace of a circular track can be expected. The water was raised to a 38 cubic metre water tank. At a later stage, the water was channeled from the mountains through aqueducts, thereby avoiding the use of further machinery.

The well at the Casa della Regina d'Inghilterra is very similar to the other two. Pemp was greatly puzzled by a number of deep marks in the well-shaft, not all of which [fig. 24 and p. 41] fitted his reconstruction.

Almost the same arrangement can be seen at the Terme del Foro. Pemp gives no reconstruction of this waterworks. He believes that each waterworks held two potgarlands and that the power from the treadwheel was transmitted to the potgarlands by means of two sets of angle-gears. A sketch is given in Fig. 107.

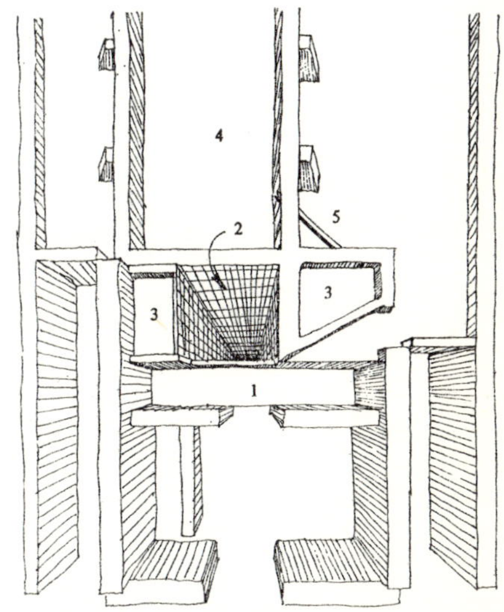

Fig. 107. Well and water tanks at the Terme Stabiane in Pompeii. No. 1, wall round treadwheel; no. 2, deep well; no. 3, two small basins; no. 4, large water tank; no. 5, gully from basin to large water tank. Machine No. (11).

Personal investigation of the three rectangular wells at Pompeii

Report

The first well is easy to find if one walks round the Terme Stabiane.

I examined the remains of the tank on the roof and the two small basins. From the southern basin, a pipe carries the water to the big tank and this pipe is, in fact, the only clear trace of the machinery, but it does not suffice to give any idea at all of even the main outlines of the structure. If there was indeed a machine here at one time, it must have been rather effective. A treadwheel powered by two men would probably have been able to fill this 38 cubic metre tank in 24 hours.

Machine No. (11)

Machine No. (12)

At the Terme del Foro there is a covered well located 23 metres to the right of the entrance and 17 metres into the ruins at right-angles to the street. This well is also rectangular and resembles that of the Terme Stabiane. The walls of the well are encrusted with lime, but a close examinations of this sediment gave no result, and I found no marks from potgarlands.

Machine No. (13)

The Casa della Regina d'Inghilterra is situated just beside the Terme Stabiane. Its correct location is Regio *VII* Ins. 13, not Regio *VII* Ins. 14. At any rate, it lies in the street between Ins. 13 and Ins. 14. I walked down the right-hand side of this street from the corner of the Strada dell'Abbondanza, and found the well hidden by bramble thickets.

There was room for a big treadwheel in front of the well. There were no traces of an elevated tank, and it was impossible to see whether there had been one potgarland in the shaft, or two; in fact, it was impossible to see whether there had ever been a machine at all.

Machine No. (14)

According to Holger Mygind: *Die Wasserversorgung Pompejis.* [Janus Vol. 22 (Leiden 1917), p. 315], there should also have been some sort of potgarland machinery near the square water tower close by the entrance to the Foro Triangolare, Regio *VIII.* Here, I made the mistake of looking inside the tower for the machinery instead of outside or in its vicinity. All that Mygind says about the machinery is that it must have been located near the tower.

Machine No. (15)

In Regio *VIII*, Ins. 5, No. 36, there is a well in which a water-lifting machine must have operated. The well is 20 metres deep, and there are a number of holes in the lining of the shaft that may have had some connexion with the machinery. Amedeo Maiuri: *Pompei .... Pozzo e impianto idrico.* [Notizie degli Scavi di antichità, Series 8, Vol. *IV* (1950/1951), pp. 128–129]. Unfortunately, I did not find this reference until after my visit to Pompeii.

Luigi Jacono, an Italian engineer who lived very near Pompeii, has also investigated the water-lifting machines there. His investigations must have been carried out concurrently with, but independent of, those of Pemp. Jacono's investigations led to a reconstruction in the form of a model, which is now at *Il Museo della Civiltà Romana*, EPU Roma. There is a coloured photograph of this model in the illustrated work:

Tutto su Roma antica, published by Bemporad Marzocco, Firenze 1963. I myself have seen only the Danish edition of the book.

By and large, Pemp and Jacono have both reached the same solution: two potgarlands driven through several gears by a large treadwheel.

Conclusion

There are four deep wells at Pompeii, the internal dimensions of which – about 2.7 × 2 metres – justify the assumption that paternoster wheels were once used in them. Adjoining the wells there is a chamber that might have housed a treadwheel. However, there is very little real evidence in support of this assumption, the only indications being two small basins at the top and a number of very faint wear marks that definitely cannot have been caused by a potgarland.

The possibility of a gear having been used cannot be excluded, but it is more likely that the method followed by Vitruvius, in which a treadwheel and a bucket chain shared a common shaft, was utilized, (cf. p. 161).

Vignette No. 8. This Italian angle-gear noria was unknown to Vitruvius. Cesure Cesariano: Vitruvius Liber Decimus, *Como 1521 p. 170.*

f. *The tomb-painting of Alexandria*

I B.C.
Egypt

In 1960, a find was made in Alexandria that was unique in the history of water-lifting machines. It is also important for the whole history of technology, because the painting in question depicts a very complicated piece of machinery.

A sketch, drawn as a copy of my colour slide, is reproduced in Fig. 108.

Machine
No. (16)

The painting shows two oxen harnessed to a machine. The tractive force is transmitted to an upright shaft through a neck-bar. Behind the shaft there is a perpendicular wheel, provided with many spokes. The transverse lines that can just be made out on the rim must indicate the separating lines between the compartments. A column supports one end of a horizontal beam, which must be provided with a bearing for the upright shaft. The entire machine is placed on a round platform, in the brickwork of which there is an outlet. Level with the outlet there is a

Fig. 108. Painting from Alexandria. Two oxen harnessed to a sāqiya or a tābūt. The outlet can be seen between the two birds. This painting has been dated to about the time of the birth of Christ. Machine No. (16).

pond on which a duck is swimming. The oxen are being led round by a boy playing the Pan-pipes.

Although this work of art is by no means a blueprint, it does provide us with valuable information.

The oxen are moving round in the "right" direction, *i.e.* against the sun. The use of a single, long neck-bar is, however, a poor solution because if one of the oxen pulls more than the other, the latter can simply follow without pulling. This unequal distribution of the work load can be avoided by having the animals side by side under a joint yoke, with the trace fixed at the middle. See Fig. 109.

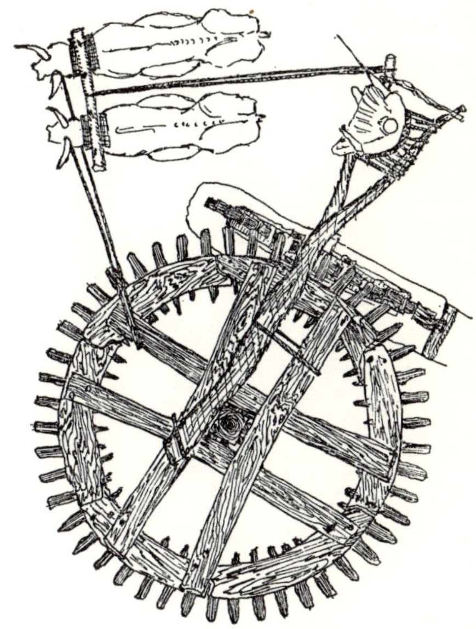

Fig. 109. Sāqiya drawn by two oxen.

Unfortunately, we cannot see in the painting how the rotation of the upright shaft is transmitted to the compartmented wheel on the horizontal shaft, but we are here in the happy position of knowing that there is only one possibility, *viz.* through the angle-gear. There are waterwheels in Egypt in which the horizontal cog-wheel is located in a pit, and if this were the case here, we should, of course, be unable to see the cog-wheel.

The wheel must be of the compartmented rim type. This is the most common type of water-lifting machine around Alexandria today. By comparing the painting with Fig. 54 a, I am inclined to think that what we see to the right of the duck is a water outlet. All things considered, we must assume that the painting depicts a tābūt, see Figs. 30 and 31, and not a potgarland sāqiya.

A reproduction in colour is given in Th. Schiøler: *Øseværket, en gammel maskine, men hvor gammel*? [Naturens Verden, Vol. 46 (Ejnar Munksgaard, Copenhagen 1963), p. 211]. The same picture is included in the Swedish edition of the above journal [Naturens Värld, No. 2 (Hälsingborg 1964), p. 51]. A better reproduction, also in colour and of greater value because it was taken before the painting was restored, is to be found in Henry Riad: *Tomb Paintings from the Necropolis of Alexandria.* [Archaeology, A Magazine dealing with the Antiquity of the World, Vol. 17, No. 3 (N.Y. 1964), pp. 169–172 and on cover].

The picture has been dated to the period about the beginning of the Christian era. According to P. H. von Blanckenhagen: *The Odyssey Frieze.* [Mitteilungen des Deutschen Archaeologischen Instituts, Roemische Abteilung, Vol. 70 (Heidelberg 1963), p. 146]:

> In my opinion the date cannot be earlier than the end of the first century B.C., and may be later. In no way can this picture demonstrate the Alexandrinian origin of landscape painting.

The painting is today in the Graeco-Roman Museum in Alexandria, room No. 15.

3. OTHER ARCHAEOLOGICAL FINDS

Although the present book is supposed to be devoted to machines that raise water when supplied with power, there are so many points of great similarity between such machines and the converse machine, *viz.* the waterwheel or gravity wheel, that it is not out of place to mention a few examples of the latter here.

V–VI A.D.
Athens

The Athenian agora is located by the carpark below the Acropolis in Athens, and here there are the ruins of a watermill that once powered a quern. At first glance there does not seem to be any great difference between these ruins and the ruins at Ostia, but a closer investigation reveals that the two are different. One of the points of similarity is provided by the end wall of the waterwheel pit. Here, there is just the same niche as described in connexion with the Ostia wheels. During the rotation of the wheels, large quantities of water are thrown against the end wall, thereby washing the mortar out from between the bricks until, finally, these fall out.

Since the excavation and surveying work in about 1936, a number of bricks have been removed from the site, unfortunately including one that showed signs of wear from the cog-wheel. Luckily, however, there are slight traces in the brick underneath (1969). The traces are so faint that investigators might hardly be prepared to accept that they originated from a wheel, but the brick does, at any rate, show a concave wearing mark. The chord has a length of about 220 millimetres and a maximum depth of curvature of 9 millimetres, which implies a diameter of between 1000 to 1200 millimetres for a cog-wheel. As far as I know, this is the only trace of a cog-wheel that has been preserved from antiquity.

Arthur W. Parsons: *A Roman Water-Mill in the Athenian Agora.* [Hesperia, Vol. *V* (1936), pp. 70–90] and Anon.: *Excavation in the Athenian Agora.* Picture Book No. 7. (American School of Classical Studies at Athens, Princeton 1961).

V A.D.
Istanbul

Opposite Hagia Sophia there is a small museum, The Museum of Mosaics, in which there is a mosaic depicting a big waterwheel working beside a high building. In the literature this is described as the waterwheel for a mill, but it could just as well be a hydraulic noria lifting water to the top floor of a house. G. Brett: *A Byzantine Water-Mill.* [Antiquity, Vol. 13 (1939), pp. 354–356].

III A.D.
Rome

Another example, where only the details can reveal whether the device in question is a waterwheel or a water-lifting machine, is provided by a wall painting in Rome, a picture of which is being sought by Lynn White, Jr., p. 82. I have succeeded in finding such a photograph in Bibliotheca Hertziana (Max Planck Institut), 28 Via Gregoriana, Rome. The photograph, which is in black and white, bears No. 15161. The original fresco is still to be found in the catacombs near the Church of S. Agnese (cimitero maggiore).

On the basis of this photograph I have made the sketch shown in

Fig. 110, but this gives only the main outlines of the painting. A colour photograph would reveal far more details – for instance, that there are brush strokes at the top of the wheel, which might possibly depict falling water. According to A. Profumo, these strokes do indeed represent falling water. [Nuovo bulletino di archeologia cristiana, Vol. *XXIII* (1917), p. 108]. Other pictures in this catacomb have been dated to the third century A.D. See P. Testini: *Le catacombe e gli antichi cimiteri cristiani in Roma.* (Roma cristiana, Vol. *II*). Bologna 1966, p. 163. According to Lynn White, Jr., p. 82, note 2, the fresco in question is the only antique picture we have of an overshot waterwheel. The wheel has 16 spokes and a round shaft.

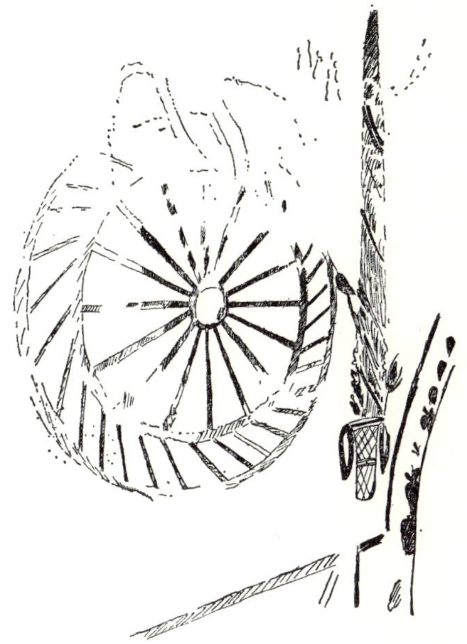

Fig. 110. Old Christian wall painting from the Catacombs at Rome. Overshot wheel for water-mill. This is the oldest known picture of such a wheel of 3rd century date.

a. *After Vitruvius*

In the foregoing, the archaeological finds in this field have been described, and these provide a rather clear picture of the water-lifting devices in use at about the time of the birth of Christ. We might therefore expect to find many references to such machines in classical literature, but unfortunately, these records are strangely silent on the subject.

To this, however, there is a single exception, namely Vitruvius: *De Architectura.* This work describes, among other things, a number of water-raising machines, although it does not mention the geared paternoster wheel.

Before studying Vitruvius, we should first mention a number of authors and finds from the first to the fifth century A.D.

A.D. *473*
Syria

M. de Vogüé: *Syrie centrale, architecture civile* . . . Paris 1865, Vol. *I*, p. 95.

Plate 57 shows a drawing of The Bath of Sağillā. Here we can see two square columns about 1.7 metres in height, which are located on top of a 16 × 2 × 4 metre underground water tank. Other drawings by Vogüé – although not this one – have been reproduced in Publications of the Princeton University Archeological Expeditions to Syria, Division *II*, Section *B*, ed. H. C. Butler, pp. 118–123, together with a number of photographs. Vogüé describes the columns as supports *d'un appareil*, while Butler uses the term *a pump*. From this *pump*, the water is led to:

> . . . . an exterior conduit, or canal, cut in the upper surface of a projecting course of stone at the level of the window sills . . . .

As in the case of Quṣayr al ʿAmra (cf. p. 92), the water is led into the building at a height of about 1.7 metres.

In plate 55 in Vogüé [p. 120 in Butler], the colums at letter *P* are designated by two small squares between which there is a small black dot. The dot must symbolize a hole down to the water tank. In such case, the hole must be about 25 centimetres in diameter, and taking all things into consideration, the *pump* appears to be the ruins of a windlass.

J. B. Ward Perkins *et al.*: *The Hunting Baths at Lepcis Magna.* [Archaeologia or Miscellaneous Tracts .... Vol. 93 (1949), p. 174]:

I–IV A.D.
Libya

> ... cisterns .... a projecting stone bracket carried a pulley, or similar device, for raising the water ....

These few words may well conceal a complete piece of machinery.

At Tolmetta (Cyrenaica), Libya, there are two pairs of storage basins with a well in the middle. The possibility cannot be excluded that there was once some type of water-lifting device here, although hardly a sāqiya. Gennaro Pesce: *Il "Palazzo delle Colonne" in Tolemaide di Cirenaica*, Roma 1950, p. 54, figs. 53, 75; Plates *I* and *XII*, No. 37 and 38.

G. R. H. Wright: *Cyrene.* [The Illustrated London News, July 16, 1955, p. 111]. (Cyrene lies between Benghazi and Tobruk in Libya). Wright shows here a sketch of a 1.7 × 1.7 metre well shaft. In 1963, Mr. Wright wrote to me as follows:

> ... I suppose the restricted size of the shaft suggests to me something more in the nature of a simple windlass, however, what you envisage [a sāqiya] may well be possible.

II–IV A.D.
Egypt

(The Excavation of Medinet Habu, Vol. *V*) U. Hölscher: *Post-Ramessid Remains.* [The University of Chicago Oriental Institute Publication, Vol. 66]. On page 40 we read:

> The first well to be mentioned is that built by Ramses *III* in *L* 11. (Court *E*,)[128] [See Vol. *III*, p. 68 and fig. 42 (Publication Vol. 54)]. Its shaft was constructed of large stone blocks up to the Ramessid floor level, then of baked bricks (31 × 14.5 × 6 centimetres) ..... up to Roman level ..... finally in Coptic period ...
> The stairway by means of which water had originally been carried up in jugs was later not used and became filled up. Instead, the water must have been lifted through the shaft by means of a mechanical device, whose nature can no longer be determined. We noted only that in the immediate vicinity of the shaft lay a granite beam about 3 metres long which was roughly rounded like a column shaft. It could have served to span the shaft opening, in which case the lifting device might have been similar to a modern sakieh. The water which was raised presumably flowed in a fairly steady stream.

R. Mond and O. H. Myers: *Temples of Armant, I.* London 1940, p. 38:

> .... Armant el Heyt .... No remains of the wheel itself were found ..... In the Coptic town a well with a stairway was excavated.

G. Roeder: *Hermopolis 1929–1939*, Hildesheim 1959, pp. 199–200, § 10*e*:

> ... Meter 150–160: in römischer oder koptischer Zeit .... In

Roman or Coptic times a 6 metre long trench was made for a mechanical lifting device, possibly a sakiya ....

301
Rome

The Edict of Diocletian. See Tenney Frank: *Rome and Italy of the Empire.* (An Economic Survey of Ancient Rome, Vol. *V*, Baltimore 1940, pp. 365–367).

In this edict there is a long list of prices for materials and finished products, *e.g.* timber for wagons and the prices of two- and four-wheeled wagons, a threshing sledge and a plough. These are followed by:

line 52	horse mill with stones	1500 dr.
line 53	ass mill	1250 dr.
line 54	water mill	2000 dr.
line 55	hand mill	250 dr.

but no water-lifting machines.

Late *II* A.D.

Artemidorus Daldianus: *Onirocriticus I*, 50.

Here we find a sentence which is appropriate to the type of machine found at Ostia, *i.e.* the treadwheel:

> ... he does not move but walks. He was sentenced to the *antlia.*

Antlia means water-lifting wheel or just irrigation wheel. We find this term used in a number of other texts.

120
Rome

Suetonius: *Tiberius*, Chapter 51:

> .... they were sentenced to the *antlia* ....

In these two examples we can conclude that men were working the *antlia.*

II A.D.
Syria

Several books on the history of technology include a reproduction of the oldest known mosaic of a hydraulic noria, see, for instance, Needham, plate *CCXXXIV*, facing p. 359 (cf. p. 42).

The mosaic was found in Northern Syria, and is now in the garden of the National Museum in Damascus. Besides the wheel, it shows the stepped walling, which constitutes the support for one end of the shaft. This mosaic was first published in Bulletin des Musées royaux d'art et d'histoire, 3 series, Vol. 5, p. 5, fig. 5.

I–II A.D.
Italy

G. Ucelli: *Le navi di Nemi*, 2nd ed. Roma 1950, pp. 181–184. Here there is a reconstruction of a hand-operated bucket chain with flywheel, which is based on finds from two different ships from the first or second century. Lynn White, pp. 106 and 168, calls this reconstruction a reliable twentieth century design .... but which is archaeologically a fantasy. I share this view; Lynn White does not believe in the flywheel and I do not believe in the bucket chain, which derives from a mediaeval manuscript.

In the next text we are in doubt as to whether it is a man or the stream that turns the wheel.

A.D. *38–102*

M. Valerius Martialis: *Epigrammaton Libri, IX*, 18.4:

> ... the curved *antlia* lifts the heavy water ....

A.D. *23–79*

C. Plinius Secundus: *Historia Naturalis, XIX*, 60:

> .... There is no doubt that the farm must have a garden and that this must be irrigated from a nearby river or from a well equipped with a pulley (rota) or pneumatic devices (organisue pneumaticis) [I doubt whether this should be a piston pump], or a šādūf (tollenonum haustu) .....

Some authors see a sāqiya in this text, but the translation of the term *rota* as pulley is probably correct.

I A.D.
Rome

Samuel Ball Platner and Thomas Ashby: *A Topographical Dictionary of Ancient Rome*. Oxford U.P. 1929, p. 195:

> .... At Transistoria .... a very large latrine which has been wrongly thought to be the machinery chamber of a hydraulic lift, which would, it is supposed, have worked in a shaft over 120 feet deep found not far off.

In order to be quite certain, I visited the site in question. I could find nothing in the way of a machine room or remains that might indicate any kind of machinery.

I A.D.
Egypt

W. M. Flinders Petrie: *Naukratis* Part *I*. (The Egypt Exploration Fund, London 1886, plate *XVII*, 13). See Fig. 111. It is difficult to decide from this tiny sketch of a terracotta fragment whether it really depicts a treadwheel, although it shows a number of steps and the feet of man. Nevertheless, the fragment is of some value if we compare it with line 149 in *P. Lond. 1177* (cf. p. 124), in which we find:

> .... the labourers *pro rata* at the ladders 7 .... dr. 30 ob. 1.

Fig. 111. Terracotta fragment of a treadwheel from the first century A.D.

As these lines are included in the sections on wages for ox-drivers at the sāqiyas, it must be assumed that the steps refer to some other machines, *viz.* treadwheels.

A common feature of the archaeological finds mentioned in this chapter is the considerable uncertainty of the authors, which is due to the fact that such machines are so rare in archaeology.

Vitruvius: *De Architectura Libri Decem.*

As a sequel to our digression let us recapitualte the archaeological finds that have been dated to the period of Vitruvius. These comprise the bucket chains with treadwheels at Pompeii, the geared water-lifting machine at Tuna al Ğabal and the painting in Alexandria.

About B.C. *20*
Rome

The text of Vitruvius was originally accompanied by drawings, but these have been lost and it is therefore necessary to rely solely on the text, which has been translated many times. Here, we shall use a translation by A. G. Drachmann: *The Mechanical Technology of Greek and Roman Antiquity*. Copenhagen 1963, pp. 150–157.

Vitruvius, Book X, Chapter 4, Section 1:

> Now I shall explain about the implements that are invented for the lifting of the water, how they are made in various sorts. And first I shall speak of the drum. This does not lift the water high, but it pumps out quickly (expeditissime) a great amount.

In the last sentence, Vitruvius reveals a well developed technical insight. He knows that the two main characteristics of a pump are its head and its discharge, and we get the impression that he is thinking in terms of power, which is the product of these two. The clue that leads us to believe that he is thinking along the right lines is the word "quickly". He does not write that it pumps out a great quantity, but rather: ". . it pumps a great quantity out quickly", a slight difference in wording, the insight of which we may admire.

> An axle made on the lathe or by compasses, with its ends iron-plated, having round it for the rest of its length a drum made of boards joined together, is placed in posts that have iron plates in them under the heads of the axle. In the hollow of its drum are placed eight dividing walls that touch the axle and the farthest round of the drum, separating equal compartments of the drum.

Section 2. Round its end are fastened lids leaving openings of half a foot [about 15 centimetres] for the water to come in. Also along the axle pigeon-holes are bored into each compartment at one end. It is tarred like a ship; it is turned by treading men; and taking in through the openings that are in the ends of the drum, it gives out through the pigeon-holes along the axle, where a wooden trough is placed with a gutter joined to it. In this way it provides a great amount of water for irrigating gardens or managing saltpans.

Among the details we note particularly the pigeon-holes and the wooden trough, which correspond to the type of construction shown in Fig. 13.

The next structure corresponds exactly to the wheels at Ostia and to the wooden machines found in the Spanish and Portuguese mines.

Book X, Chapter 4, Section 3:

> But if it has to be lifted higher, the same method is adapted (?) like

> this: Let there be a wheel round an axle, of such a size that it fits the height for which there is need. Round the outer side of the wheel are fastened square boxes made tight with tar and wax. So when the wheel is turned by treading men, the full boxes, carried up on top, and returning to the bottom, all by themselves pour out into a container what they have carried up of water.

There is such a wheel at the museum in Cracow, Poland. It was found in the salt mines at Wieliczka and has been dated to the 17th century. The wheel has a diameter of about 5 metres and is provided with boxes as described by Vitruvius.

An improved version – the compartmented wheel – has the boxes built into the rim.

There is a comprehensive literature on these wheels, which were used in Roman mines. Caro Baroja (1954), pp. 38–40, with 3 drawings. Forbes, Vol. *VII*, pp. 215–217. For a new and valuable contribution we are indebted to George C. Boon and Colin Williams: *The Dolaucothi Drainage Wheel.* [The Journal of Roman Studies, Vol. *LVI* (1966), pp. 122–127]. This article mentions the remains of a wheel found in Wales, which is now in the National Museum of Wales at Cardiff. In the British Museum, Gallery of Greek and Roman Life, there is another wheel, which came from Minas de Ríotinto, (37°45′N, 6°36′W). Finally, I have seen one wheel at the Musée du Conservatoire national des arts et métiers (Paris), which is not mentioned above or in the following. It has a diameter of about 5 metres and is provided with a hub of about 1 metre in diameter. The wheel is rather well preserved and has 2 × 28 spokes. It is labelled: "7.269 Roue hydraulique élévatoire provenant des mines de San Domingo, *V*e siècle. Don de Deligny. Entré 1864".

This Deligny must be identical with Ernest Deligny 1820–1898, a French mining engineer, who was engaged in Spanish mining for many years and also interested himself in Roman mining. See S. G. Checkland: *The Mines of Tharsis. Roman, French and British Enterprise in Spain.* London 1967, pp. 64–75.

At the Museo de Ríotinto there is yet another wheel, and a picture of this is to be found in J. M. Luzón: *Los sistemas de desagüe en minas romanas del suroeste peninsular.* [Archivo español de arqueología, Vol. 41 (Madrid 1968), pp. 101–120].

Luzón concludes his article by noting that there must still be many traces of Roman technology in the Ríotinto mines that are as yet unknown to us and, further, that the British Museum possesses a scale drawing of the Ríotinto wheel in the Gallery of Greek and Roman Life.

The next passage has been quoted earlier but will be repeated here, in its proper context. It deals with the type of machine that operated in Pompeii with bucket chain and treadwheel.

Book X, Chapter 4, Section 4:

> But if it is to be delivered in higher places still, round the axle of the wheel there is placed a double iron chain let down to the lowest level, with hanging bronze buckets on it, each holding a *congius* (3.3 litres). So the turning of the wheel, by winding the chain round the axle, lifts the buckets up on top, which, when they are borne up over the axle, are forced to turn over and pour out into a container as much water as they have lifted out.

Vitruvius points out *that a double iron chain* must be used; this is because a single chain would twist and the buckets would get caught up against the bucket-chain wheel. There are many examples of bronze buckets from Roman times in museums, but none of those that I have seen has any connexion whatsoever with bucket chains. A number of special bronze buckets have been found in the Spanish and Portuguese mines. Oliver Davies: *Roman Mines in Europe*, Oxford 1935, p. 25. Luis de Albuquerque e Castro: *Hallazgos romanos en la mina* "*Do Fojo das Pombas*". [Archivo español de arqueología, Vol. 35 (1962), Nos. 105 and 106, pp. 166–176]. There is some doubt as to whether these buckets have been used in bucket chains, although the following paper tends to substantiate such an assumption. Guillermo Gossé: *Las minas y el arte minero de España en la Antigüedad.* [Ampurias. Revista de arqueología, prehistoria y etnología *IV* (Barcelona 1942), pp. 56–57].

In the copper mines of Sortiel Coronada in the Province of Huelva, there is a 1.5 metre diameter well with a depth of 6–7 metres, which it is thought once contained a bucket chain because of the three large Archimedean screws found at the top of the well. If there is to be any reasonable relationship at all between the quantity of water supplied by the screws and the depth of the well, one can only conclude that there must once have been a bucket chain in use here.

The last machine described by Vitruvius in this context is the hydraulic noria. As noted earlier, this has been mentioned by other classical authors and is also to be found in a mosaic.

Book *X*, Chapter 5:

> Wheels are also made in the rivers on the same principles as those mentioned above. Round their rims are placed paddles which, when they are pushed by the power of the river, go forward and force the wheel to turn and so, lifting by boxes and carrying up on high, being turned by the power of the river itself, without the treading of workers, they perform what is needed for use.

Immediately after describing the hydraulic noria, Vitruvius passes to the geared watermill because this too is driven by water power, but he then returns to water-lifting machines.

In a long passage [Book *X*, Chapter 6], he describes the construction of an Archimedean screw. We recognize some of the Vitruvian design details in the paper by Gossé, drawing No. *X*. It should be noted, however, that the screw construction detailed by Vitruvius can only be described as clumsy. Today (1964), an Archimedean screw is made as follows:

First, about 70 blocks measuring 35 × 65 × 430 millimetres are cut, and a 15 millimetre hole is drilled through the middle of each. The blocks are then slipped onto a 14 millimetre iron rod and arranged round this in the form of a spiral staircase. The staircase is then clad with boards nailed to the ends of the blocks. The iron rod and the blocks with cladding thus constitute a single element and not, as some believe, a screw rotating inside a tube. There are, of course, water-screws working on this principle, but they are of a much later date. The water-screw described by Vitruvius is a rigid construction.

Book *X*, Chapter 7 contains a description of the piston pump of Ktesibios. Such a pump has been found in Minas de Sotiel Coronado in Spain, and a drawing is included in the paper by Gossé, facing p. 65. This technological sensation, which Gossé calls the pearl of Hispano-Roman antiquities, is in the Museo Arqueológico Nacional, Madrid. The pump, which is a twin-piston pump, is very well preserved and is a far finer example than the two from Bolsena, which are in the British Museum. It is mentioned only in Spanish literature. There is another, slightly better drawing of it in Luzón, facing p. 118 (cf. p. 160).

Book *X*, Chapter 8 deals with the hydraulic organ. An outstanding work on this subject is Jean Perrot: *L'Orgue de ses origines hellénistiques à la fin du XIIIe siècle*. Paris 1965, but since publication, a very large mosaic has been found in Syria showing, *inter alia*, a hydraulic organ with many hitherto unknown details. No publications have yet appeared (1970) about the mosaic in the museum at Ḥamā.

As will be realized from the above, there is excellent agreement between the archaeological finds and Vitruvius's text, and it is therefore strange that he never described the sāqiya. This omission leads us to believe that he was not acquainted with it.

b. *Before Vitruvius*

B.C. *48* *Egypt*

Julius Caesar: *Bellum Alexandinum*, Chapter 6:

> ..... After cutting off the water-tanks and closing all the parts of the city that he had occupied, he hastened to press a great amount of water in from the sea by means of wheels (rotis) and machines (machinationibus).

It is unfortunately impossible to determine what type of machine Caesar is describing.

Strabo: *Geographica. XVI.* 1.5 (738): B.C. *66–24*

> ... By the side of the stairs are water-engines (kochlias), *i.e.* Archimedean screws by means of which persons appointed expressly for the purpose are continually employed in raising water from the Euphrates into the garden.

This description relates to the Hanging Gardens of Babylon. It is obvious that Strabo was acquainted with the Archimedean screw, but it is questionable whether this ingenious invention was also known in Babylon in the 6th or 7th century B.C.

In the next passage we read about screws and also about some wheels. *XVII* 1.30 (807):

> .... Wheels (trochos) and screws (kochlias) lift the water from the Nile to a camp. One hundred and fifty men are employed in this work.

These words may also refer to a sāqiya, but there is insufficient information for us to decide beyond doubt.

T. Lucretius Carus: *De Rerum Natura, V.* 515: B.C. *95*

> ... wheels and pots are turned by the river ...

This sentence illustrates the eternal rotation of the heavens, and it must be identical with what, in this book, is called a hydraulic noria. Most of the technical terms used in the passages quoted have been treated by M. G. Bruno: *Il lessico agricolo latino e le sue continuazioni romanze.* [Istituto Lombardo ......, Rendiconti Vol. 91, *III–IV* (Roma 1957), pp. 998–999], but the translations are far from precise.

Diodorus Siculus: *Bibliotheca Historica*: *II*, 10. The Hanging Gardens of Babylon: B.C. *100*

> ... One arch had in it certain conduits and contrivances from the platform ....

We shall revert to this when we come to the archaeological investigations made at Babylon, (cf. p. 164).

Philon of Byzantium: *Pneumatics*, Chapters 61–65: B.C. *220*

Philon's bucket chain has been dealt with earlier, in chapter 3 on Arabic illustrated manuscripts, (cf. p. 65), but besides this strange contraption, he mentions three other wheels that are turned by water, and these are certainly all of Islamic origin. See A. G. Drachmann: *Ktesibios, Philon and Heron.* Copenhagen 1948, p. 66.

III B.C. It is at this point of the chronology that the one of Ptolemaic kings enters the history of irrigation. A papyrus, (cf. p. 127), shows that he was very interested in mechanical irrigation.

V B.C. There are papyri in other languages than Greek, see, for instance, Arthur Ernest Cowley: *Aramaic Papyri of the Fifth Century* B.C. Oxford 1923. This book includes an important papyrus on shipbuilding, but none on mechanical irrigation.

VII B.C. As one of the Seven Wonders of the World, the Hanging Gardens of Babylon have naturally been the subject of a great deal of discussion, and interest has concentrated particularly on the irrigation of these elevated gardens. Besides the above-mentioned passage from Strabo and Diodorus, there are reports by Robert Koldewey: *Das wieder erstehende Babylon*, Leipzig 1913, p. 92:

> . . . In one of the western chambers there is a well that differs completely from all the known wells in Babylon and other places in the world of antiquity. This well consists of three adjacent shafts, a square one in the middle, and two rectangular ones on each side. I interpret this system as part of a mechanical waterwheel . . . with bucket chain, the wheel being driven by means of a horse walking round. Such an arrangement, which is also found in the region today, is called a *dolab* . . .

It is true that in the part of Babylon designated the Hanging Gardens, there are three well shafts, but these are so narrow that a bucket chain cannot pass up and down. There is, however, the possibility that one part of the rope can go down one of the shafts and up another. In this case, the shafts would have to end in a big, common well chamber, but the wells have not yet been excavated (1969), so it is impossible to ascertain whether this is the case, and I can only conclude with Koldewey that:

> Eine definitive Lösung kann nur die Ausgrabung bringen.

Egypt (Fig. 112) All the egyptologists, except one, maintain that the sãqiya did not exist in ancient Egypt. The exception is W. Spiegelberg: *Zwei Ausdrücke der Bewässerung*. [Zeitschrift für ägyptische Sprache . . . . . . Vol. 53, p. 113].

In answer to this paper, H. Schäfer writes in the following volume of the same journal, pp. 140–141, that as long as there are no archaeological finds that point to the sãqiya, it would be wiser to use the translation, water-lifting apparatus *Schöpfwerke* instead of waterwheels *Schöpfräder*. Similar conclusions are reached by H. E. Winlock and W. E. Crum: *The Monastery of Epiphanius at Thebes*, N.Y. 1926, p. 65, and Ugo Monneret de Villard: *La Nubia romana*. (Pubblicazioni dell' Istituto per l'Oriente, Roma 1941, p. 44). It is difficult to find evidence of the non-existence of the sãqiya, but there is an article by Henry De Morant: *Secrets of the "Sacret Lakes" at Ancient temples*: A long-hidden structural prob-

Fig. 112. This picture must not be taken as evidence that the sāqiya was known in ancient Egypt. The picture dates from about 1920 and the sāqiya was invented about the time of Christ.

lem solved. [The Illustrated London News, March 19th, 1938, p. 498–499], which states, among other things, that the holy lakes near the temple of Amun at Karnak and the temple of Hathor at Dandarah were not provided with water by sāqiyas or šādūfs but from underground conduits during the inundation of the Nile.

Mesopotamia
B.C. *1100*

This domain has been investigated by Jørgen Laessøe: *Reflexions on Modern and Ancient Oriental Water Works.* [Journal of Cuneiform Studies, Vol. 7 (New Haven 1953), pp. 5–26]. On page 11, the Fifth Tablet of the Babylonian Lexical Series is mentioned, and here there is a term that is interpreted as a sāqiya:

> ... But in view of the extreme brevity (no more than six lines) with which this apparatus is treated, it seems doubtful that a machine as complex as the sāqiya can be intended, when it is remembered that an implement so relatively simple as the plough occupies sixty-two lines of the same tablet.

On p. 22, we read:

> ... References to water-wheels in ancient Mesopotamia, found in handbooks and popular accounts, are for the most part based on the false assumption that the Akkadian equivalent of the logogram GIŠ.APIN was *narṭabu* and denotes an instrument for watering ("instrument for making moist").

Another reference from the Neo-Assyrian period – about B.C. 800 – that should be mentioned here is C. H. W. Johns: *An Assyrian Doomsday Book or Liber Censualis of the District Round Ḥarran: In the Seventh Century* B.C., Leipzig 1901, pp. 18–20 and 40:

> Something like half of the Ḥarran census was under irrigation. The larger machines were operated by two, four, six or even eight oxen.

In these and a few other examples, Professor Læssøe points out that the word for *plough* (Akkadian epinnu) should in all cases be substituted for waterwheel and irrigation machine.

There is one passage of particular interest, which tells of the replacement of a number of šādūfs by other devices. The only question is the nature of the new design that was to replace the old, familiar one. It should be remembered here that the passage comes from Assyria, where the groundwater table lies at such a low level that a single šādūf would not suffice to raise the water to the surface. In the text, which is from the 7th century B.C., it is stressed that the new machine consists of ropes and chains made of bronze, and that the beam is of date-palm wood, which indicates that the beam must have had a considerable length. We are further informed that the machine stands over a well. Professor Læssøe writes of this apparatus:

> I suspect a winch [windlass or a great pulley]; but the technical term for this apparatus is not used.

The term applied to describe the beam appears to mean *high* (*i.e.* elevated) *beam*, and, bearing this in mind, I visualize a different apparatus from that proposed by Professor Læssøe.

In India there exists a gigantic šādūf, which is called a picottah. This is used at very deep wells, and the horizontal lever is located 6–8 metres above ground level. The machine is operated by three men, two of whom walk backwards and forwards along the lever, thereby acting as mobile counterweights. A ladder is provided for the use of the labourers when they have to reach *the high* (elevated) *beam*. As this is a heavy-duty machine, short chains are necessary to carry the enormous leather sacks in which the water is raised.

There is only one term, maiāltū, that is used in relation to wheels and oxen and it might therefore mean a sāqiya, but here Læssøe refers to Armas Salonen: *Die Landfahrzeuge des alten Mesopotamien.* [Annales academiae scientiarum Fennicae, Series B, Vol. 72,3 (Helsinki 1951), p. 36]. He states that maiāltū is undoubtedly a wagon.

The last reference I have found on this subject is also by Armas Salonen: *Agricultura Mesopotamia nach sumerisch-akkadischen Quellen.* [Annales academiae scientiarum Fennicae, Series B, Vol. 149 (Helsinki 1968), p. 376]. In the chapter on oxen, reference is made to a translation with the wording: *boeufs de machines pour l'irrigation.* M. Lambert. [Revue d'assyriologie et d'archéologie orientale, Vol. 59 (1965), p. 66]. However, Salonen asserts that the translation is incorrect.

As a result of his investigations, Læssøe writes as follows on the question of the sāqiya:

> I consider it unlikely that any reference to the sāqiya will appear in ancient Mesopotamian sources.

In his opinion, we should turn our attention to Alexandria, "where it seems plausible to assume that the sāqiya was invented."

Vignette No. 9. The well is a popular motif in old Christian art but there is not much technical information to glean for the history of technology. The twelfth century. Painted ceiling panel, Cappella Palatina, Palermo, Italy. See Richard Ettinghausen: Treasures of Asia, Arab Painting. *Albert Skira 1962, p. 127.*

CHAPTER 8

Historical Conclusion

We occasionally encounter the assertion that the sāqiya was known in ancient Mesopotamia or pharaonic Egypt, and this is usually based on the belief that this machine is so simple that even a man with no special qualifications can design it. Nothing could be further from the truth; the waterwheel is an extremely complicated machine, and its inventors must have possessed a unique imagination and a thorough knowledge of contemporary technological possibilities. There are no indications that this machine goes back to the ancient cultures – it is quite simply an invention of the engineers of classical antiquity.

Its introduction was due to Archimedes (B.C. 287–212), whose mathematics and hydrostatics bear the stamp of that unique imagination which is a prerequisite for the creation of new machines. He himself invented the water-screw, which equals the sāqiya in technical ingenuity.

There was, at the same time, a receptiveness to new ideas about methods of raising water in Egypt. This is demonstrated by a papyrus addressed to one of the Ptolemies. Nevertheless, although there are thousands of Greek papyri from the second and first centuries B.C., not one of them provides convincing proof of the existence of the sāqiya.

The sāqiya consists of three components: a paternoster chain, a right-angle gear and a circular track. Investigations at Pompeii indicate that the paternoster chain was known about the year B.C. 100, and the circular track was used in oil-mills and grinding mills, but the question of when the big right-angle gear was invented remains unanswered. The sources at our disposal today point to the 1st century B.C. The most important source in this respect is the sāqiya system at Tuna al Ğabal. Here we can see the cog-wheel pit for the vertical cog-wheel. Vitruvius also describes a right-angle gear in Book *X*, Chapter 5, Section 2. (A. G. Drachmann: *The Mechanical Technology* .... 1963, pp. 151–152).

Heron, who lived in about A.D. 60, wrote fairly extensively on right-angle gears in his *Mechanics*, and by reading the passages in question, we come to realize just how difficult it was – for the scholars of that time – to describe the theory of the gearwheel. (A. G. Drachmann (1963), pp. 81–86). We see that Heron understood the principle involved, but could not express himself correctly because he lacked the concepts of angular velocity and torque.

However, besides its theoretical aspect, the right-angle gear has important technological implications. The people of that time undoubtedly had small toy gears, but it takes a certain amount of courage to produce one on a large scale. Innumerable attempts undoubtedly preceded the final form.

At first, the sāqiya was a luxury article. The size of the water tanks at Tuna al Ǧabal make it obvious that only small quantities of water were supplied. The painting in Alexandria shows a tābūt (sāqiya) in a luxurous garden. Not until *P. Lond. 1177*, from A.D. 113, is a large sāqiya-system mentioned in a waterworks for a town. By about the year 250, sāqiyas were in use on big estates.

At many places in Egypt, the sāqiyas are on 24-hour operation for 100 days at a time, and a stoppage of just a few days can ruin a crop. So Allāh is asked to protect the machine and the animals against misfortune. In other words, the sāqiya has to have a high degree of reliability. The peasant who, in the Greek papyrus, asks the carpenter to hurry with the dispatch of the machine "because we need it" gives us the impression that the crop has already been ruined. It is obvious to all that the sāqiya has many advantages, but it is a complicated and sensitive piece of machinery consisting of more than 200 different components. Far simpler and therefore also more reliable is the šādūf, which has only about 10 parts. In most places, the šādūfs work in pairs, so that even if one breaks down, the other will save the crop. In *P. Lond. 1177*, mention is made of two sāqiyas working in pairs for 118 days and in this period, one of the machines breaks down three times.

There is a further example of the problems relating to reliability to which attention should be drawn. At Abu Mina, the sāqiya was replaced by a treadwheel. The reason for this may well have to be sought in the greater reliability of the latter machine.

It is clear from the above that the sāqiya had many competitors, and I believe that it was not until the introduction of the earthenware pot and the pawl mechanism in the 4th or 5th century that it became popular.

The sāqiya was also known in Roman Syria, possibly at a rather early date, though the only support for this assumption is a couple of pots.

I am convinced that the sāqiya was known in the Roman colonies along the North African seaboard, but I have not yet succeeded in finding any references to such machines in the literature.

Subsequent dissemination is difficult to trace, but in the 8th century we encounter the sāqiya in the Syrian desert at the summer pavilion of Quṣayr al ʿAmra in Jordan. There is a miniature painting showing a special type of waterwheel from Baghdad, which can be dated to the 12th century. From this city, the machine found its way to India.

In view of the strong ties that existed between Damascus and Valencia in the field of irrigation methods, it is probable that the noria came to Spain from Syria rather than from Morocco. See Fig. 113. During the excavations at Ḥamā, a noria-pot with a waisted middle was unearthed. See Fig. 74. Almost the same type of pot can be seen at the museum in

Fig. 113. Noria from southern Morocco, 1965.

Málaga. See Fig. 73. It is also significant that there is a type of noria located about 60 kilometres north of Ḥamā, which has much in common with the Spanish prototype.

The theoretical treatment of water-raising machines also has a history of its own, which is, in fact, the development of the idea of mechanical efficiency.

The story commences with Vitruvius, who shows by his choice of words that he understood that head and discharge are the two quantities that best describe the various types of machines. He further gives us the impression that he had a vague idea of the relationship between these two quantities and the power. However, two other quantities are also required for an investigation of the efficiency of a machine, one of these being the concept of angular velocity, and the other the torque. In this field, too, we note the uncertainty of the classical authors (Heron and Pappos). When they write of gearwheels, they think in terms of statics regardless of the fact that the problem is one of dynamics.

Much later, we realize from a passage in al Yaʿqubī (dec. 891) that the concept of perpetual-motion machines had still not been defined. People thought along other lines at that time; machines were classified into two groups, *viz.* the group powered by animals or men and the group to which the waterwheel belongs, powered by the perpetual flow of the river.

At the end of the 11th century, an Indian astronomer designed a number of perpetual-motion machines – on paper. These were copied by the Arabs, who deliberately made both drawings and text unintelligible. In doing so, they adhered scrupulously to the Indian text, which reads: ". . this is a mystery; if plainly described, it would be generally intelligible in the world." Ebenezer Burgess: *Translation of the SŪRYA-SIDDHĀNTA, a Textbook of Hindu Astronomy.* [Journal of the Americna Oriental Society, Vol. 6 (1890), p. 405].

Ğazarī followed slightly different paths, producing marvellous toys.

One of his machines is a sãqiya, apparently powered by a wooden ox, but in fact powered by a concealed impact wheel.

The Arabic manuscripts on perpetual-motion machines form the basis for the European literature in this field. The literature on perpetual-motion machines and on water-lifting machines is extensive, but in both cases, the structures presented are far removed from the world of reality. An excellent example of this mixture of fantasy and fact is provided by Peter Mundy. During a visit to Warsaw he was told of a waterwheel that turned by itself at the same time as it lifted water, but by the time he saw it, the machine had been altered so that it was now powered by 12 slaves (1643).

Concurrently with the increasing industrialization at the end of the 18th century, the theoretical treatment of the machines finally got on the right track. Mr. Townsend's investigations of the Spanish noria show that he had a correct insight into the problem. He concentrated on the machine's loss, and described a primitive investigation of its efficiency, although he had his doubts about the results obtained. In 1737, Bélidor calculated the efficiency of a sãqiya.

This interplay between real machines with a low efficiency on the one hand, and theoretical designs with a somewhat too high efficiency on the other, one day acquired great importance.

On the basis of the innumerable attempts that had been made through the ages to construct a perpetual-motion machine, the French Academy in 1775 reached the by no means self-evident conclusion that it was impossible to contrive such a device.

In 1842, Sadi Carnot took a major step forward. For the purpose of describing his cyclic process, he used both a real machine and an abstract model. *i.e.* a theoretical design. Carnot wrote: . . . that whereas the conditions concerning machines "driven by water, air, or animal power can be completely analysed by mechanical theory, a similar theory for the study of heat engines is conspicuously lacking". D. S. L. Cardwell: *Power Technologies and the Advance of Science, 1700–1825* [Technology and Culture, Vol. 6 (1965), p. 189]. In order to explain the conditions relating to the heat-machine, he then introduced a machine with no loss, which is, in reality, a theoretical concept.

Conclusion:

The sãqiya has been in use for almost two thousand years, and we might imagine that its days are numbered now that modern pumps can do the work faster and sometimes also more cheaply. The truth of the matter is, however, that the sãqiya and the noria are increasing in number; they are old machines with which the peasant is completely familiar, while the motor pump is a temperamental piece of machinery that makes unprecedented demands on specialized knowledge or a well-filled wallet.

Chronology for the Geared Water-Lifting Wheel

2000	FAO publication on water-lifting machines. An economic analysis.
1900	Ingenieur Barrois describes the Egyptian sãqiya. Napoleon's scientists investigate the sãqiya.
1800	Mr. Townsend gives a technical description of the noria. Frederic Norden draws an Egyptian sãqiya. Bélidor is the first to calculate the effciency of a waterwheel.
1700	Stanisław Solski of Poland has a couple of drawings of sãqiyas. Imaginary waterwheels in Böckler: *Theatrum Machinarum Novum.*
1600	Waterwheels in Italy. Persian miniatures with waterwheels. Georgius Agricola: a few imaginary waterwheels.
1500	Juanelo Turriano's ms. in Madrid has a couple of waterwheels. Jacopo Mariano has a good drawing of an indoor waterwheel.
1400	The Citadel and Joseph's Well at Cairo.
1300	Spanish and Syrian waterwheel pots with the same profile. Ğazarī's illuminated ms. on, *inter alia*, toy waterwheels.
1200	Ibn al ʿAwwām: the Spanish noria.
1100	Miniature from Baghdad. An excellent waterwheel construction. Ibn Sīda describes the sãqiya.
1000	Arabic papyrus.

900 Al Balāḏurī describes, perhaps, a waterwheel in Iraq.
Arabic papyrus.
A Coptic legend mentioning a sāqiya at Alexandria.

800 Al Kindī: sāqiyas with cisterns in Egypt.
Quṣayr al ʿAmra: remains of a sāqiya system.

700 Greek papyrus telling of the little gearwheel.
Coptic papyrus telling of a wheel with 40 cogs.

600 Many Greek sāqiya papyri.
Abu Mina: remains of a sāqiya system.

500 Abu Mina, remains of another sāqiya system.
Greek papyrus mentioning a wheel with 36 cogs.

400 Dated waterwheel pots.

300 Greek papyrus mentioning oxen for irrigation.
Greek papyrus: farm accounts mentioning the sāqiya.

200 Greek papyrus mentioning several sāqiyas with ox-drivers.

100 Heron discusses the gear at great length.

0 Painting of a water-lifting machine (tābūt) in Alexandria.
Tuna al Ǧabal: well preserved ruins of a sāqiya system.
Vitruvius describes several water-lifting machines.

100

200

Philon.
Interest in mechanical irrigation in Egypt.

300 Archimedes, inventor of the Archimedean screw.

Geographical Distribution

1. River Douro, Portugal
2. Ciudad Real, la Mancha, Spain
3. Málaga, Spain
4. Costa Brava, Spain
5. Majorca, near Soller, Spain
6. Ibiza, near the town of Ibiza, Spain
7. Rabat, Morocco
8. Albenga, near Genoa, Italy
9. Sardinia, Italy
10. Naples, Italy
11. Bari, Italy
12. Mistra Bay, Kalbara Valley, Malta
13. Argos, Greece
14. Chios, near the town of Chios, Greece
15. Rhodes, near the town of Trianda
16. Yediküllе, near Istanbul, Turkey
17. Cyprus
18. Aleppo, Syria
19. Dayr az Zawr, Syria
20. Beirut, Lebanon
21. Sidon, Lebanon
22. Gaza
23. The Nile Delta
24. Al Fayyūm, Egypt
25. Luxor, Egypt
26. Dongola, Sudan
27. Blue Nile, Sudan
28. Mosul, Iraq
29. Baghdad, Iraq
30. Hala, east of the Indus
31. Delhi and Agra, India
32. The River Amu Darya

I. Fez, Morocco
II. Ḥamā, Syria
III. Fayyūm, Egypt
IV. Southern Jugoslavia
V. Spain

Fig. 114. Geographical distribution of geared and ungeared norias.

Resumé på dansk

Vandløftningshjul i den romerske og islamitiske verden

Denne bogs mål formuleres bedst som et spørgsmål: »Hvornår og hvor blev det første gearede øseværk konstrueret?« Det samme spørgsmål stillede jeg mig selv for 20 år siden, da jeg så et vandløftningshjul (noria) på Ibiza.

Inden man forsøger at besvare et sådant spørgsmål, må man kende denne maskine til bunds. Et sådant kendskab opnår man bedst ved at se den i marken. Først når man har opmålt og tegnet flere maskiner, har man mulighed for at tolke de historiske kilder.

Som supplement til arbejdet i marken er det af værdi at kende først den mere nærliggende litteratur og derefter den mere specielle. I denne bog er der lagt særlig vægt på den faglitteratur, der omhandler det romerske og islamitiske verdensrige.

Desværre har det været nødvendigt at begrænse emnet. Det omfatter således kun den gearede konstruktion (sāqiya); men på den anden side, hvor det har været nødvendigt, er andre vandløftningsmaskiner medtaget.

1) Den klassificering, der er benyttet her, er hentet fra en afhandling om vandløftningsapparater i Portugal. Først deler den vandløftningshjulene i to hovedgrupper. 1) Hjul uden gear. 2) Hjul med gear. Den anden gruppe – som er denne bogs hovedemne – deles derefter i tre typer. Disse underinddeles igen, hvorved der bliver tale om 12 forskellige typer. De nævnte typer bliver derefter gennemgået; men for at undgå gentagelser, er det kun den spanske noria og den ægyptiske sāqiya, som bliver behandlet til bunds. I dette resumé skal der kun omtales et par konstruktionsdetaljer, som man sjældent kan se på et fotografi.

Øseværker skal således være forsynet med en pal, som forhindrer maskinen i at dreje tilbage. Endvidere er der på den spanske noria anbragt et par ledestænger, som forhindrer krukketovet i at løbe ud til siden.

Virkningsgraden er en størrelse, der har betydning for bedømmelsen af et øseværks effektivitet. For den spanske noria er den ca. 0,6. Denne værdi tåler sammenligning med moderne pumper.

2) I de følgende kapitler bevæger vi os tilbage i tiden for på den måde at danne os et billede af de forskellige tiders syn på øseværker. Her har jeg udelukkende holdt mig til de rent tekniske synspunkter især omkring begrebet virkningsgrad. Derimod har jeg ikke undersøgt maskinens betydning for landbrugets økonomi.

En af de første, der har behandlet den spanske noria ud fra et teknisk synspunkt, er en englænder, som i perioden 1786–1787 foretog en rejse gennem Frankrig og Spanien. Han omtaler nogle konstruktionsfejl, og forklarer derefter, hvorledes disse kunne ændres for derved at opnå en bedre økonomi.

Bélidor's værk om vandløftningsmaskiner fra omkring 1740 er værdifuldt, fordi det er skrevet af en ingeniør. Han omtaler et stort øseværksanlæg i Rochefort, og beregner virkningsgraden af dette anlæg. Det er første gang, vi møder en sådan beregning. Igennem middelalderen finder vi nu en række bøger og manuskripter, som bærer præg af forfatternes manglende kendskab til disse maskiner, men som til gengæld er i besiddelse af en uhæmmet fantasi. Enkelte af disse tegnebordskonstruktioner er udstyret med lodder og hæmværker i lighed med tårnure, en idé som også findes i ældre arabiske manuskripter.

Af langt større værdi er et par miniature-malerier med motiver fra Indien. Disse viser realistiske øseværker, som ikke afviger fra dem, vi kender i dag.

3) I den arabiske litteratur findes der en del beskrivelser af vandløftningsmaskiner; men de omhandler især de store hjul i Ḥamā, som var berømte i hele den islamitiske verden. Der findes dog nogle enkelte steder i den koptiske litteratur, hvor sāqiyaen er nævnt.

Særlig interessant er et par illustrerede islamitiske manuskripter, som kun beskæftiger sig med øseværkernes teknik. I den sammenhæng spiller Philon fra Byzants en stor rolle. I de fleste bøger, som beskæftiger sig med teknikkens historie, findes der en gengivelse af hans spandekæde. Det er imidlertid et spørgsmål, om denne konstruktion skyldes Philon. I så fald skulle et af de vigtigste elementer i øseværket, nemlig spandekæden, gå tilbage til det tredie århundrede før Kristi fødsel. Mere sandsynligt er det, at dette afsnit hos Philon er føjet til af andre, fordi pneumatik og vandløftningsapparater nu engang var sideordnede emner.

Ǧazarī's konstruktioner kan bedst beskrives som mageløse maskiner. Emnet er grundigt behandlet af Wiedemann og Hauser; men det er lykkedes mig at finde nogle bedre tegninger i Istanbul, som har færre fejl end de af Wiedemann benyttede.

Hvis man betegner Ǧazarī's konstruktioner som mageløse, må et andet manuskript med titlen *Wheels which turn by themselves* betegnes som hemmelighedsfulde maskiner. Der er tale om evighedsmaskiner, som trækker forskellige vandhjul og møller.

Begge disse afhandlinger bærer præg af indisk indflydelse.

Et mere nøgternt billede får man af datidens noria'er ved at læse Ibn al ʿAwwam's afsnit om den spanske noria. Allerede dengang – omkring år 1200 – forsøgte man at forbedre noria'en.

4) Det er nærliggende at tro, at sāqiya'en har en lang historie i Indien; men ingen af de tekster, der taler om vandløftning, peger mod en sāqiya. Derimod findes der et relief fra det 10. århundrede, der viser os et gearløst spandekædehjul.

5) Et godt billede får vi af sāqiya'en ved at gennemgå den islamitiske arkitektur. I ørkenen øst for ʿAmmān findes der endnu resterne af et jagtslot fra det 8. århundrede. Her kan man se ruinerne af et hydraulisk anlæg, der består af en brønd, hvor der til den ene side var bygget en sāqiya og til den anden side en cisterne. Af sāqiya'en er der kun oksegangen og en af de to søjler tilbage; men dette er tilstrækkeligt til, at vi kan danne os et billede af den oprindelige konstruktion.

6) Et stort arkæologisk materiale, som på en fortræffelig måde fortæller os om sāqiya'ens udbredelse, er de krukker, som blev anvendt i sāqiya'ens krukketov. Det er så heldigt, at krukkeformen har holdt sig nogenlunde uændret igennem tiderne. Især i Sudan har man fundet mange af disse krukker. Her viser de, at sāqiya'en i hvert fald var kendt i det 3. århundrede.

7) Blandt de talrige papyri, som er fundet i Ægypten, findes der enkelte, som rummer en omtale af sāqiya'en. Den første, der skal nævnes her, er en koptisk papyrus, der taler om et stort tandhjul med 40 tænder. Fra samme periode findes der en græsk tekst med ordvalget *det store og det lille hjul*. Det betyder det store og lille tandhjul i sāqiya'en. Det bedste skriftlige vidnesbyrd om sāqiya'en er ca. 200 år ældre nemlig fra år 440. Her beder en bonde godsejeren om et nyt tandhjul med 36 tænder. Regnskabslister fra de store godser viser også, at sāqiya'en var benyttet omkring år 250. Vi får at vide, at en tømrer opstiller et tandhjul og tilhugger tænder til maskinen. Også byerne havde brug for vandløftning. I en regnskabsliste fra et vandværk hører vi om maskiner, hvor der til hver er knyttet 3 oksedrivere. Der er god grund til at antage, at der her er tale om sāqiya'er, selv om materialet er for sparsomt til med sikkerhed at kunne afgøre det.

Der findes nogle papyri, der er ældre end denne, og som taler om kunstvandingsmaskiner; men ingen af dem peger mod sāqiya'en. Derimod er det sikkert, at man interesserede sig for kunstvanding med maskiner. I en papyrus fra det 3. århundrede før Kristi fødsel hører vi om en mand, der beder kongen om at lytte til hans råd, for han har opfundet en maskine, som vil frelse landet, nu hvor floden ikke er steget i de sidste tre år. Desværre fortæller han os ikke noget om, efter hvilket princip denne maskine arbejder.

Vi kommer nu til de vigtigste bidrag om vandløftningshjulenes ældste historie. Disse findes i den klassiske arkæologi.

I Abu Mina i nærheden af Alexandria kan man se ruinerne af en sāqiya med højtliggende aksel. Denne blev senere afløst af et trædehjul koblet direkte til et krukkehjul med krukketov.

Går vi nu fra det 5. århundrede tilbage til det 2. eller 3. århundrede, så finder vi i Ostia en konstruktion, der kun anvender trædehjul, men hvor fælgen består af kasser, der løfter vandet. Disse hjul er meget store, et enkelt er næsten 7 meter i diameter. Kun et enkelt sted i Ostia ser det ud til, at man har benyttet sig af en krukkekæde, men af maskineriet er der intet spor.

Omkring vor tidsregnings begyndelse blev der ca. 300 km syd for Kairo bygget et stort hydraulisk anlæg. Udformningen af maskinfundamentet viser os, at sāqiya'en ikke har ændret sig i de forløbne 2000 år. I akselgangen er der dybe slidspor efter akslens slid mod murværket.

I Pompeii findes der fire brønde, hvor man i sin tid benyttede paternosterkæder; men det er ikke nemt at danne sig noget billede af det øvrige maskineri.

I museet i Alexandria findes der et maleri, der viser os to okser, der trækker et vandløftningshjul. Af billedet fremgår det, at sāqiya'en endnu omkring Kristi fødsel var på forsøgsstadiet. I stedet for at trække på samme hammel er okserne forspændt en diagonalstang.

Desværre kniber det med at supplere de førnævnte fund. Der findes et par steder i den arkæologiske litteratur, hvor der nævnes vandløftningsapparater, men det drejer sig ofte om simple brøndvinder.

Som det fremgår af det tidligere nævnte, er det lykkedes at påvise sāqiya'ens tilstedeværelse i det romerske Ægypten ved hjælp af arkæologiske fund. Man skulle derfor vente, at der var noget at hente i den store klassiske litteratur; men det er ikke tilfældet.

Heldigvis er der en enkelt undtagelse, det drejer sig om Vitruvius' bog *De Architectura.* Her fortælles om den type hjul, som senere arbejdede i Ostia. Om de trædehjul med spandekæder, som fandtes i Pompeii og muligvis i de spanske miner. De store hjul af den type som vi kender fra Ḥamā, har Vitruvius også beskrevet. Efter en gennemgang af vandmøllen hører vi om vandskruen og til sidst om vandorglet og en stempelpumpe. Alle disse maskiner møder vi igen i arkæologien. Når Vitruvius derfor ikke omtaler sāqiya'en, fristes man til at tro, at han ikke kendte den.

Jeg har forsøgt at finde noget om sāqiya'en i tiden før det første århundrede før Kristus, men uden held.

Om kunstvandingen af de hængende haver i Babylon ca. 600–700 før Kristi fødsel kan der ikke meddeles noget nyt; sporene er for svage. Vender vi os mod det store antal kileskrifttekster, som beretter om livet og matematikken i det gamle Mesopotamien, er der nogle, som taler om maskiner trukket af okser. Men maskinerne er ikke øseværker, det er plove eller vogne.

8) Den historiske konklusion er delt i to afsnit. Det første behandler – i kronologisk orden – de fund, der belyser sāqiya'ens udvikling og anvendelse. Det andet afsnit behandler udviklingen af det teoretiske begreb virkningsgraden.

Første afsnit opfatter sāqiya'en som en konstruktion bestående af tre dele: En paternosterkæde, et vinkelgear og en oksegang. Paternosterkæden og oksegangen var kendt omkring år 100 før Kristus; men intet tyder på, at kæmpetandhjulet var opfundet på dette tidspunkt. Først omkring Kristi fødsel fandt man på, at oksegangens horizontale bevægelse kunne overføres til paternosterkædens vertikale bevægelse ved hjælp af et stort vinkelgear. Ved Tuna al Ğabel findes der tydelige spor af en sāqiya, ligesom et maleri i Alexandria viser os denne maskine.

Sāqiya'en vandt kun langsomt indpas, da det var en kompliceret konstruktion, som ikke besad den samme grad af pålidelighed som f.eks. brøndvippen. Først omkring år 250 blev den for alvor indført som kunstvandingsmaskine i landbruget. Fra Alexandria bredte den sig til Sudan, Syrien, Bagdad og senere under araberne til Spanien.

Den teoretiske behandling af vandløftningsmaskinerne har også sin historie.

Den begynder med Vitruvius, som i sit ordvalg viser, at han har forstået, at løftehøjden og vandets volumenhastighed er de to størrelser, der bedst beskriver de forskellige typer. Men vil man bestemme en maskines virkningsgrad, må man tillige kende endnu to størrelser nemlig vinkelhastigheden og drejningsmomentet. I oldtiden havde man en vag fornemmelse af disse begreber, men ingenlunde en klar definition.

Af et enket citat hos en araber kan man se, at begrebet evighedsmaskine endnu ikke var defineret i det 9. århundrede.

I Indien interesserer man sig for evighedsmaskiner på papiret, og disse overtages af araberne, som kobler dem til vandløftningsapparater. Nogle af disse konstruktioner kan følges helt op i det 17. århundrede, men i det 18. århundrede kommer man ind i de rette tankebaner. I 1737 finder Bélidor virkningsgraden af et øseværk.

Disse tanker omkring begrebet virkningsgrad og skrivebordskonstruktioner kontra virkelige maskiner danner grundlaget for Carnot's kredsproces.

Sāqiya'en har nu arbejdet i små 2000 år, men intet tyder på, at det samlede antal er faldende. Tværtom ser det ud til, at det er stigende. Sāqiya'ens historie er endnu ikke afsluttet.

Vignette No. 10. Danish sketch of water-raising machinery for the Royal Navy Dock. Manuscript from 1682 in Rigsarkivet, Copenhagen. (Søetaten 1682–84 nr. 52.)

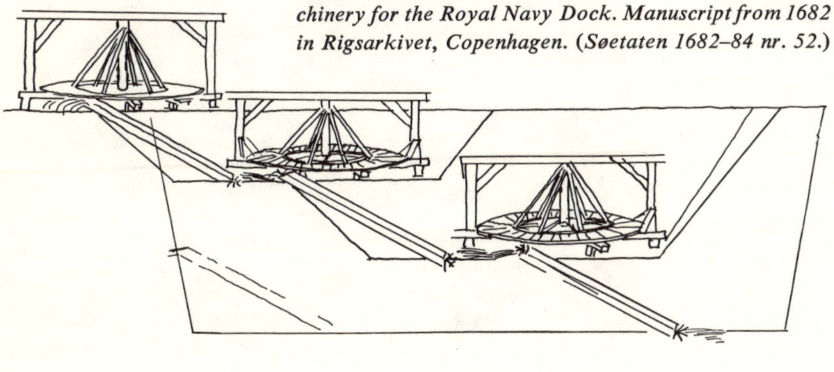

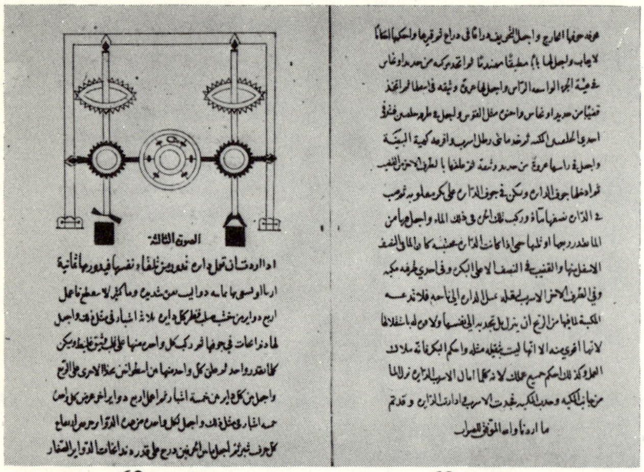

63 recto　　62 verso

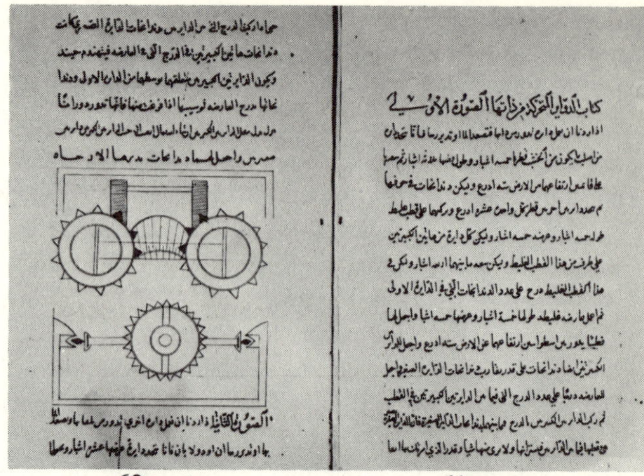

62 recto　　61 verso

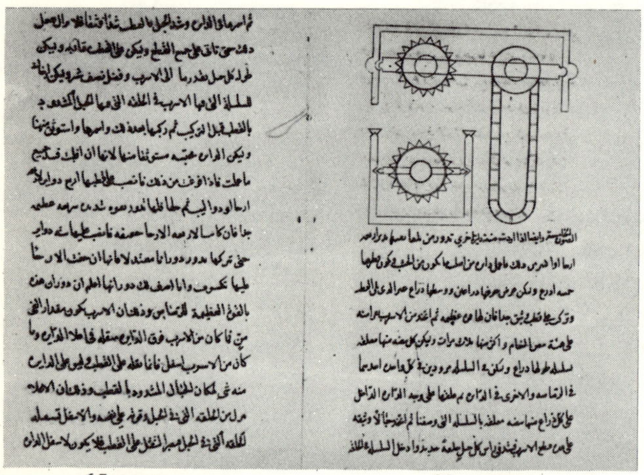

65 recto　　64 verso

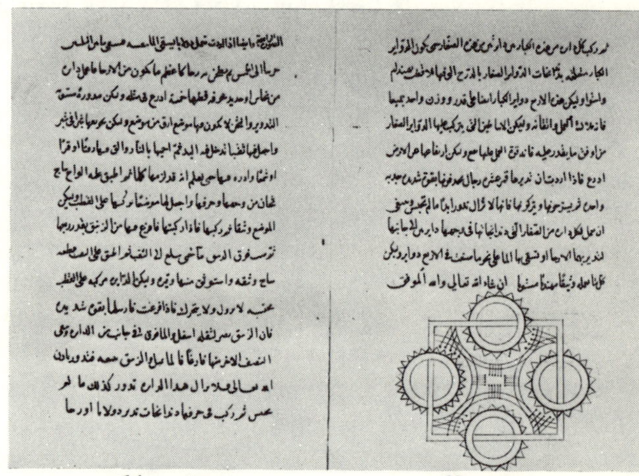

64 recto　　63 verso

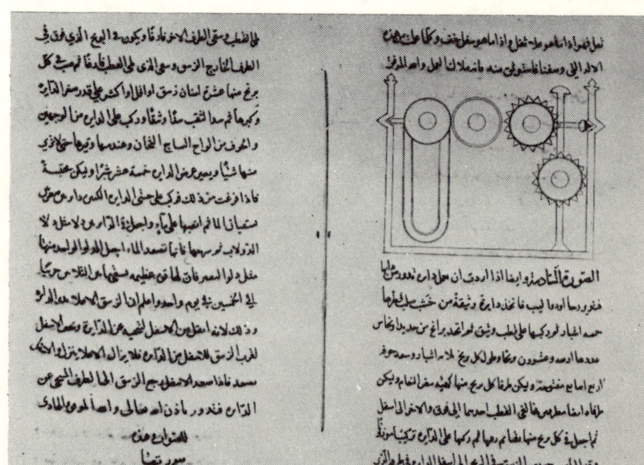

66 recto　　65 verso

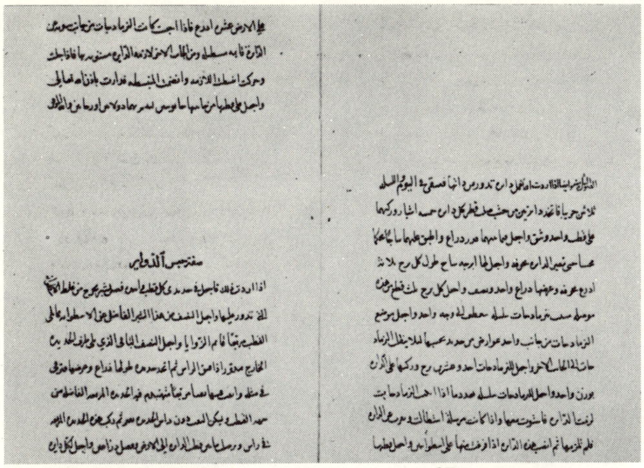

68 recto 67 verso

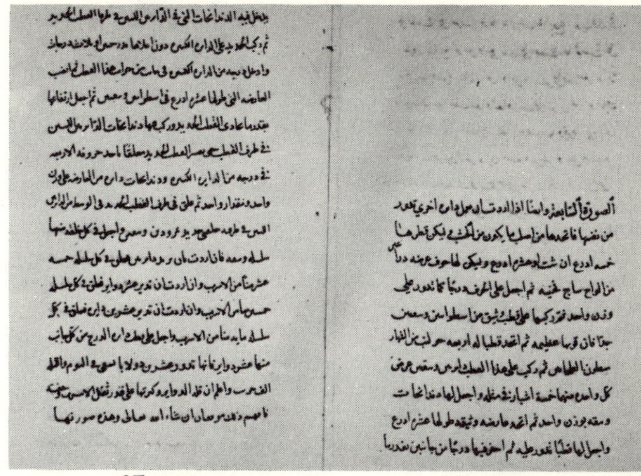

67 recto 66 verso

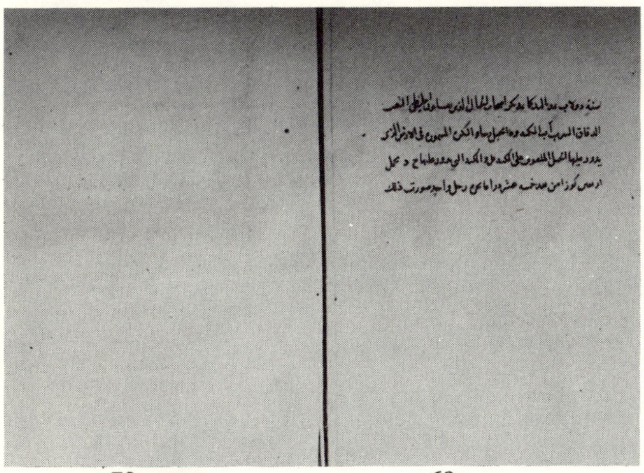

70 recto 69 verso

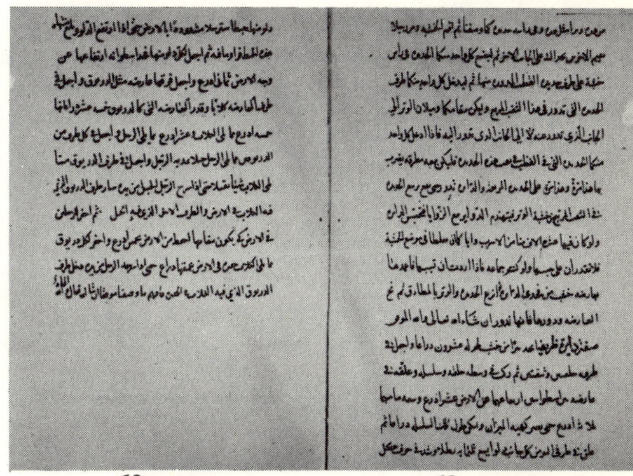

69 recto 68 verso

Plate 1. Hagia Sophia 2755 folios 61 verso to 70 recto. This is the chapter: *Wheels which turn by themselves*. It deals with perpetuum-motion machines, which in turn work some mills and water-lifting wheels. On two of the illustrations we can see the potgarlands.

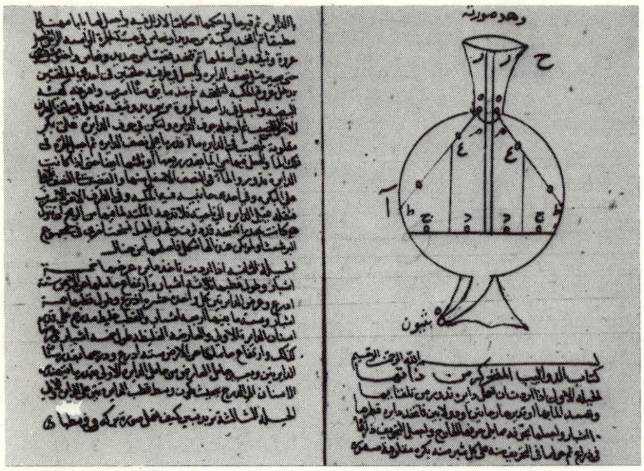

4 recto – 3 verso

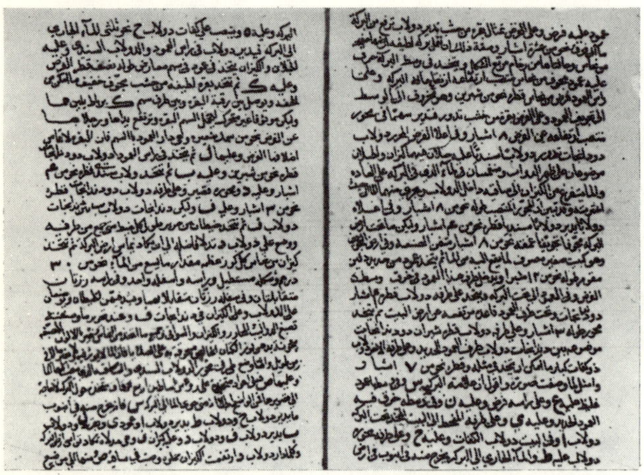

5 recto – 4 verso

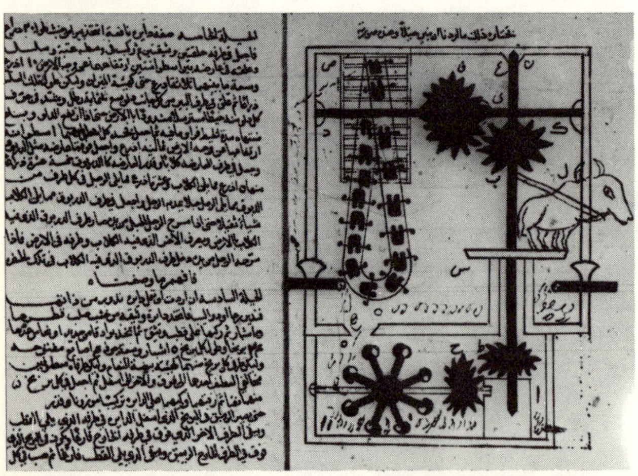

6 recto – 5 verso

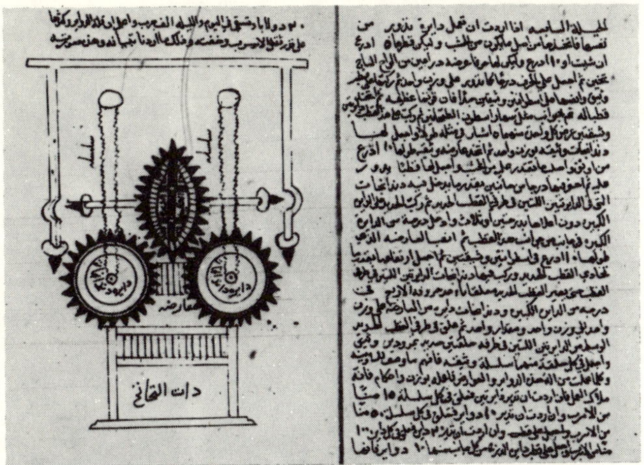

8 recto 7 verso

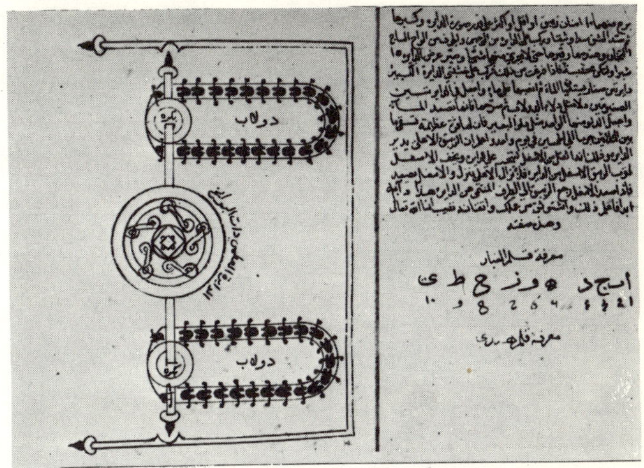

7 recto 6 verso

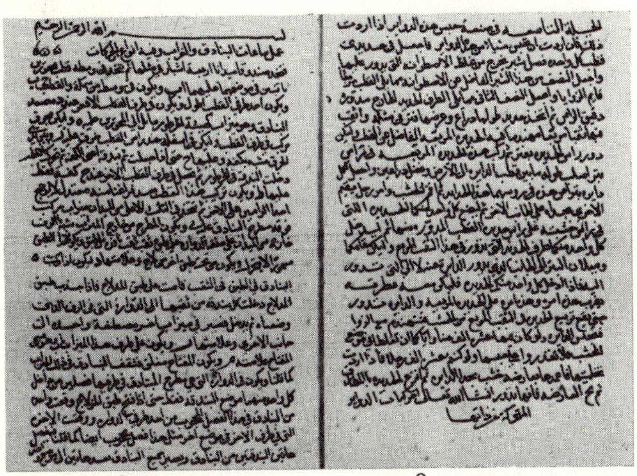

10 recto 9 verso

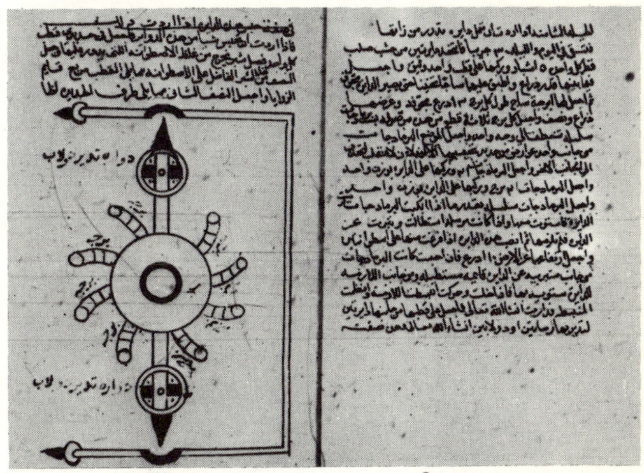

9 recto 8 verso

Plate 2. Oxford 954 = Oxford March 669, folios 3 verso to 10 recto. This is the chapter: *Wheels which turn by themselves*, but in a different version than Hagia Sophia 2755. A Ǧazarī illustration appears among the other drawings.

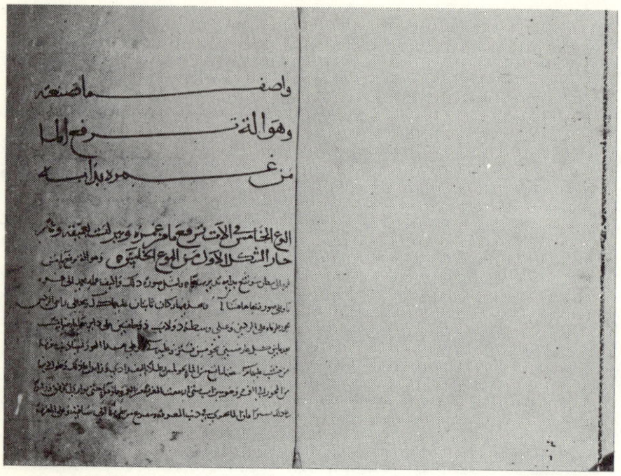

150 recto 149 verso

front folio

152 recto 151 verso

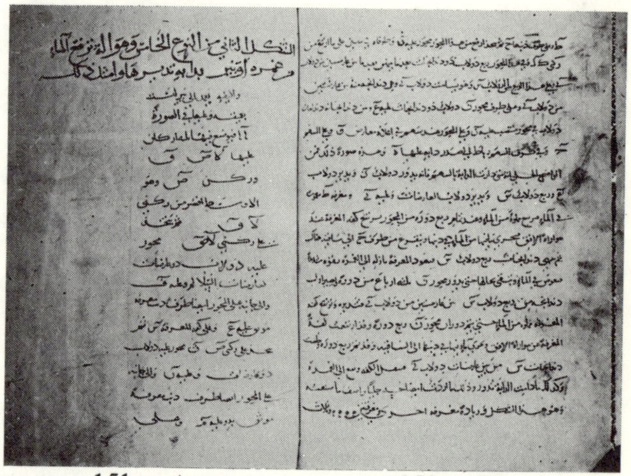

151 recto 150 verso

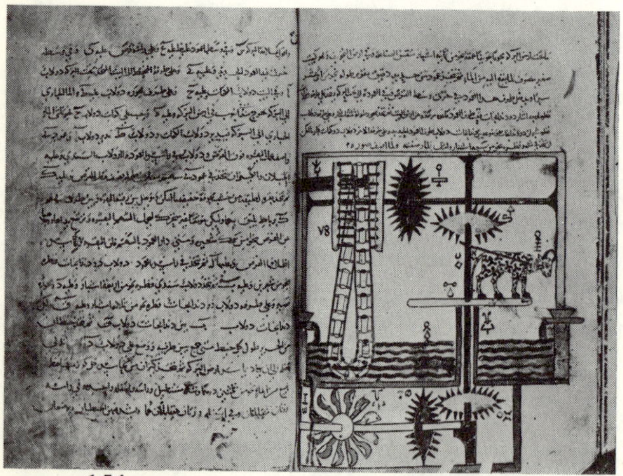

154 recto 153 verso

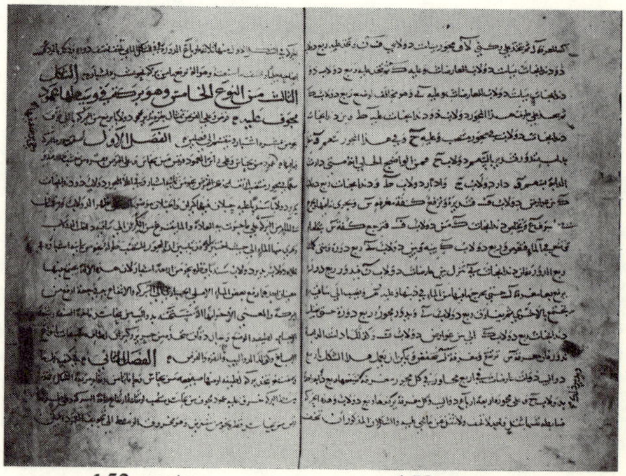

153 recto 152 verso

156 recto 155 verso

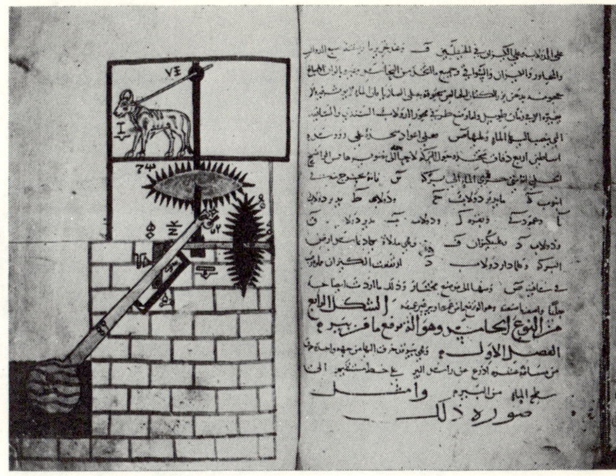

155 recto 154 verso

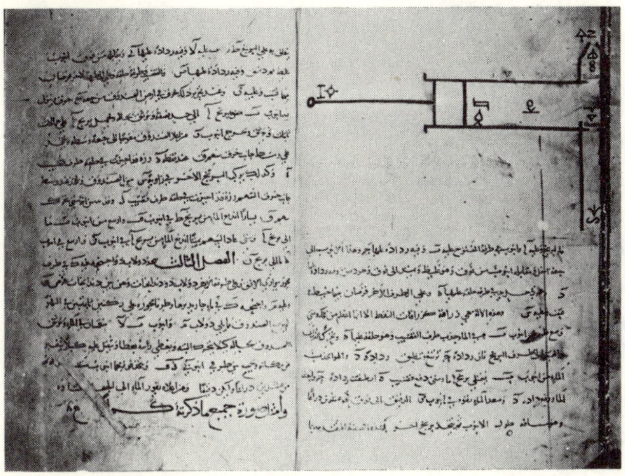

158 recto 157 verso

157 recto 156 verso

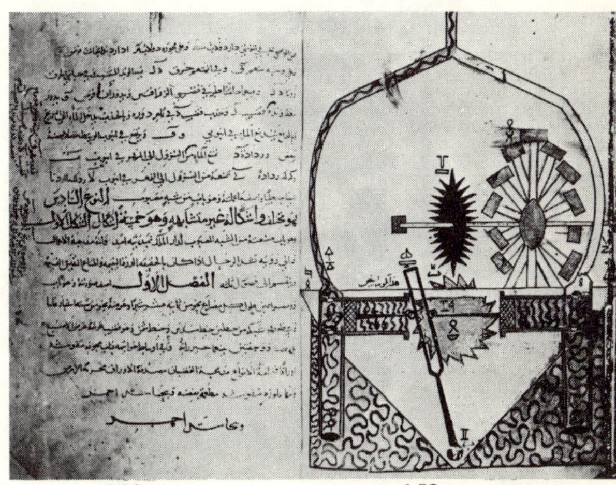

159 recto 158 verso

Plate 3. TKS Müzesi H. 414. Ğazarī.

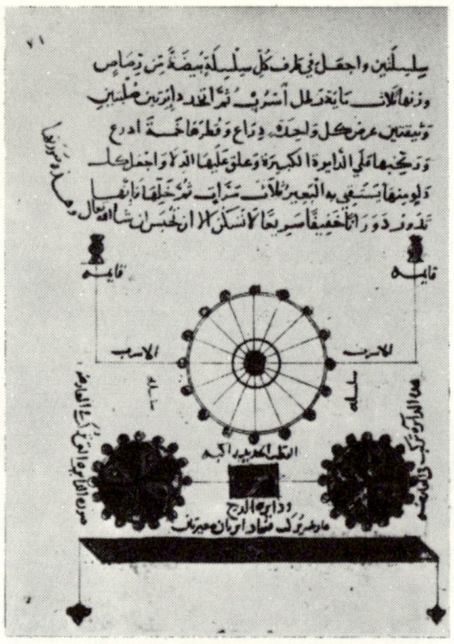

71 recto

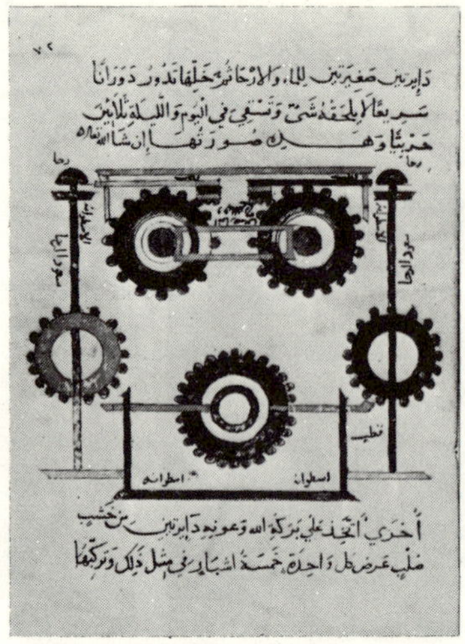

72 recto

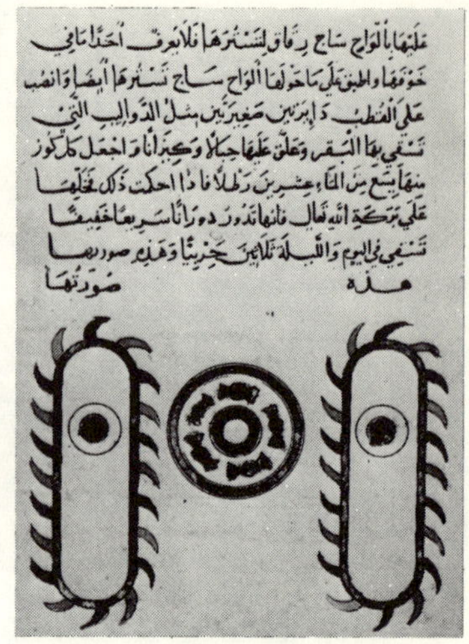

73 verso

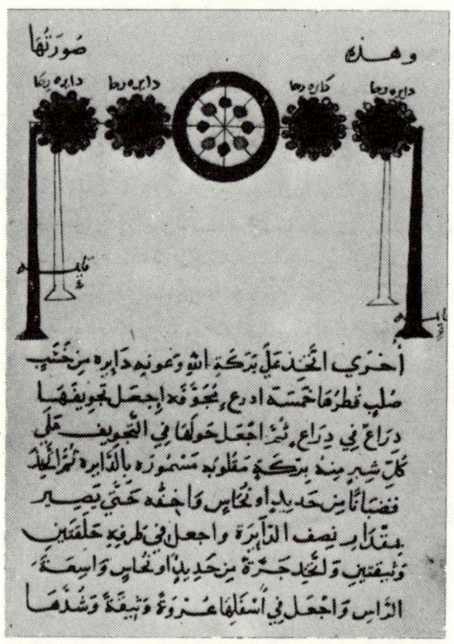

75 verso

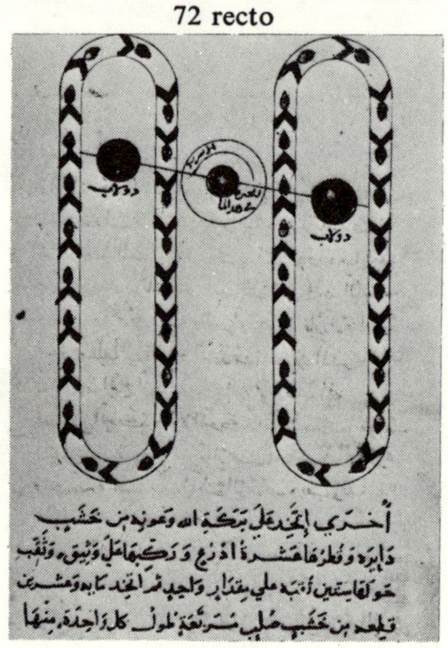

76 verso

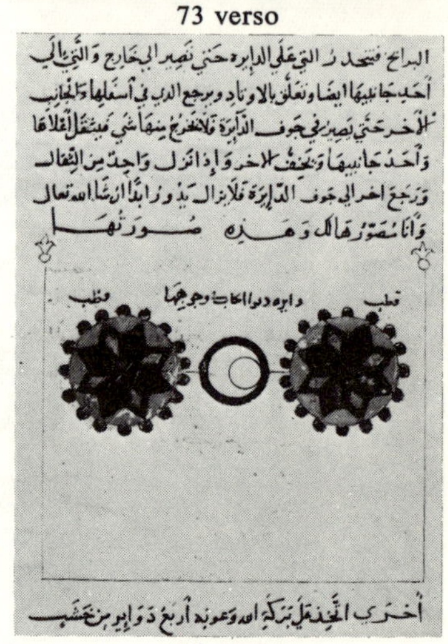

77 verso

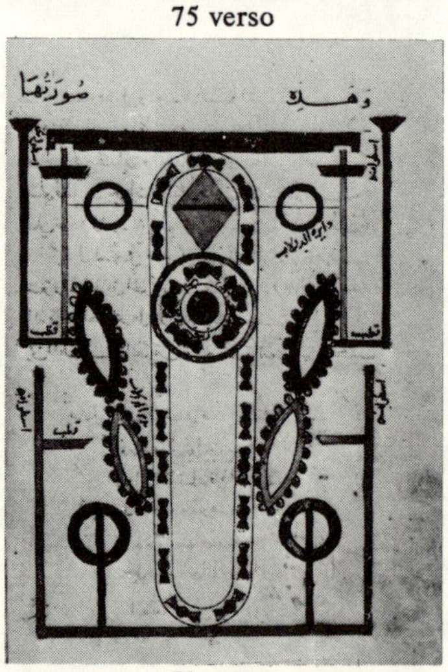

78 verso

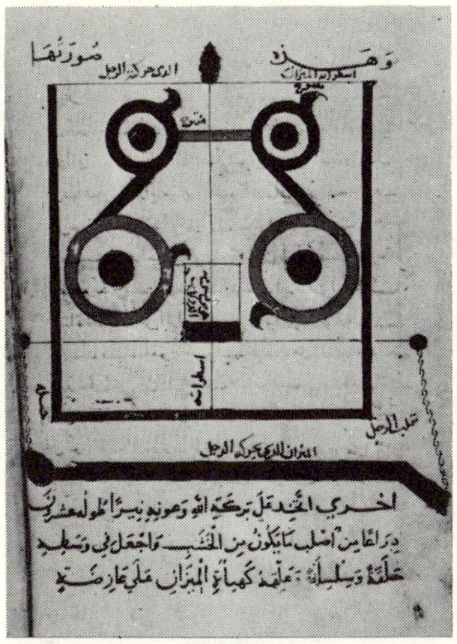

79 verso

80 verso

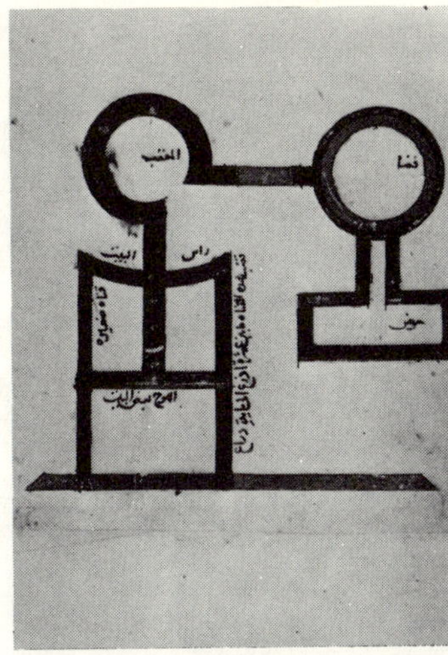

81 verso

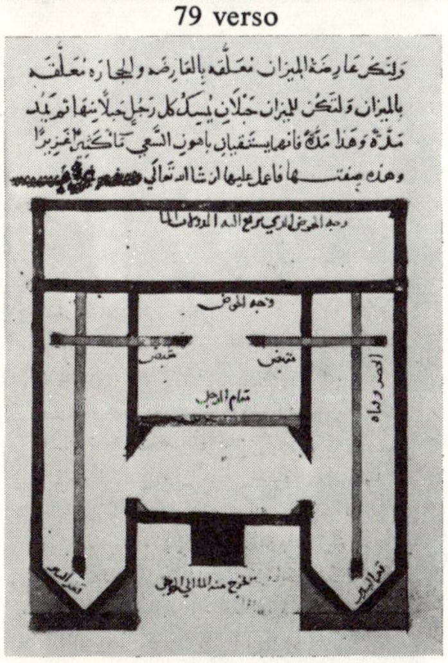

82 verso

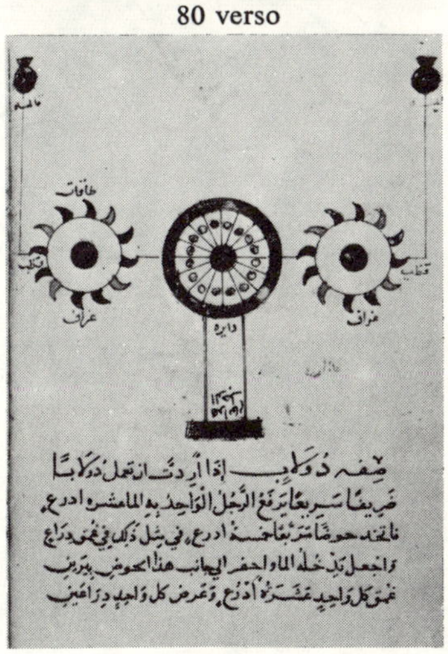

83 verso

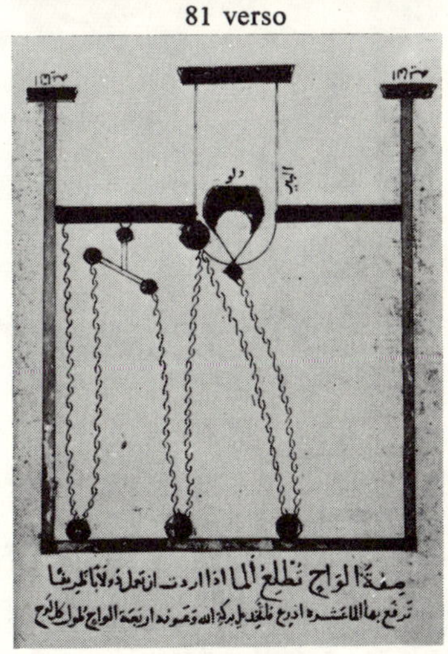

84 verso

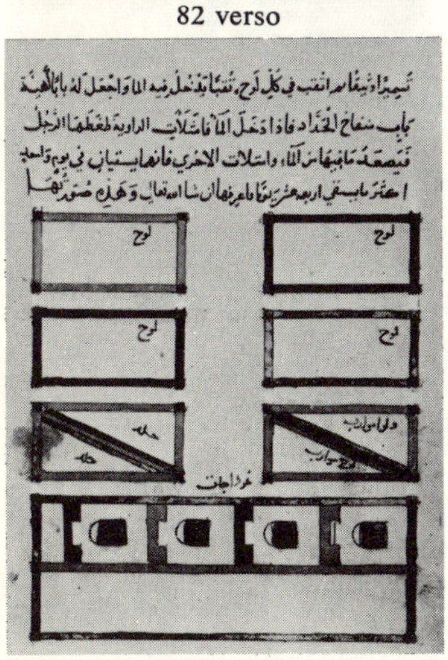

85 verso

Plate 4. Istanbul ʾAsʿad 1884. The illustrations between folios 71 recto to 85 verso. The illustration on folio 73 recto is missing.

← Plate 6. Frederic Louïs Norden's own drawing.

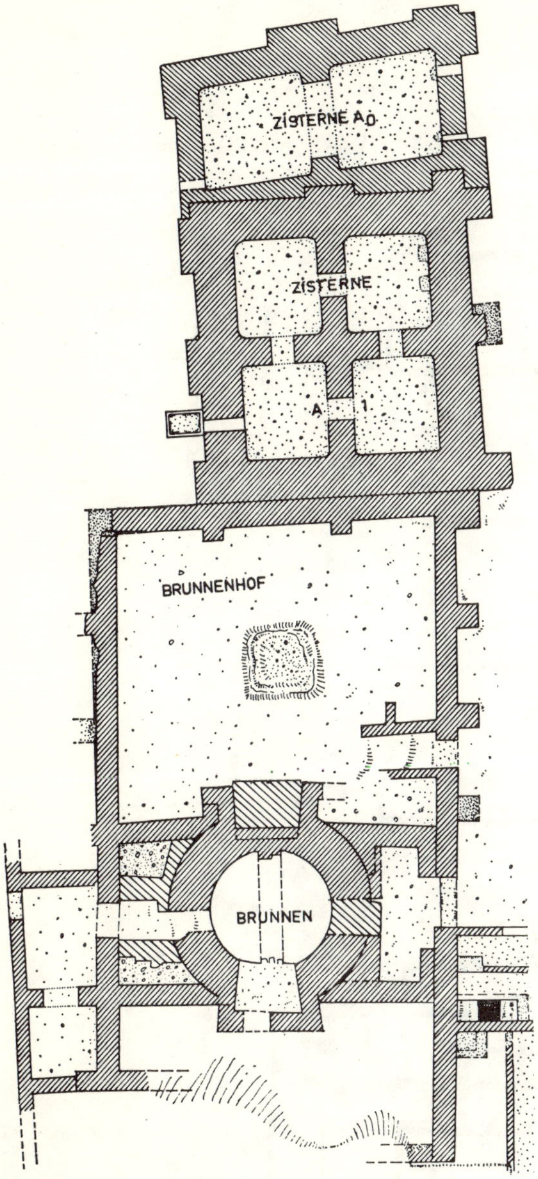

Plate 8. W. Müller-Wiener: Abu Mena 5. [Mitteliungen des Deutschen Archaeologischen Instituts Abteilung Kairo, Vol. 22 fig. 2].

← Plate 5. Copper engraving of a sāqiya in Frederic Louïs Norden: *Voyage d'Égypte et de Nubie.*

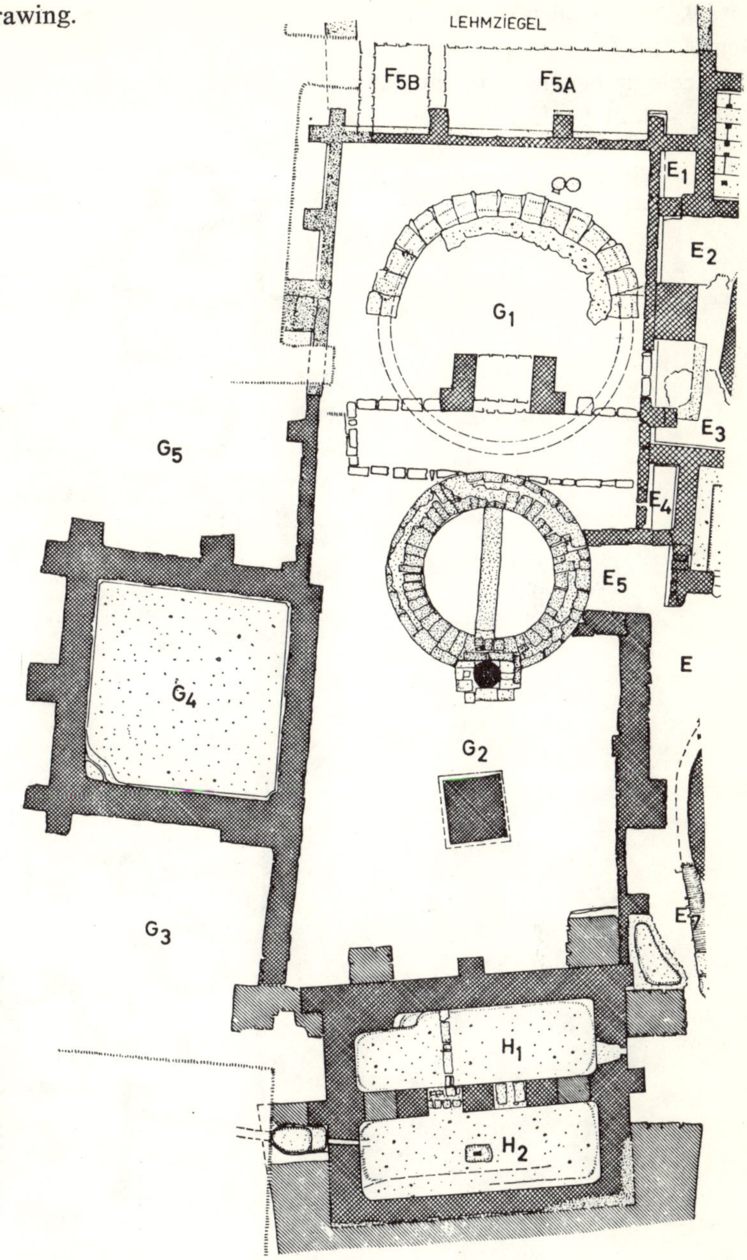

Plate 7. W. Müller-Wiener: Abu Mena 3. [Mitteilungen des Deutschen Archaeologischen Instituts Abteilung Kairo, Vol. 20 fig. 1].

Plate 9. W. Müller-Wiener: Abu Mena 4. [Mitteilungen des Deutschen Archaeologischen Instituts Abteilung Kairo, Vol. 21 figs. 2a and 2b].

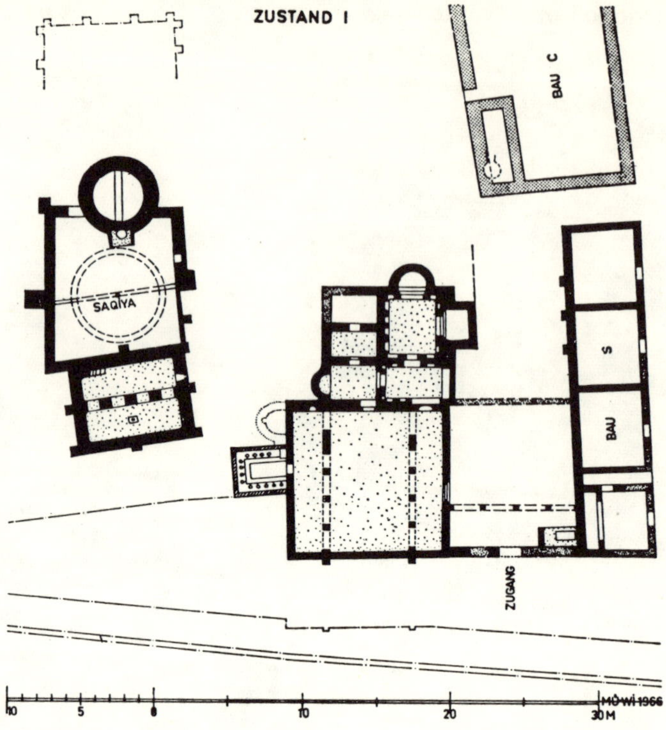

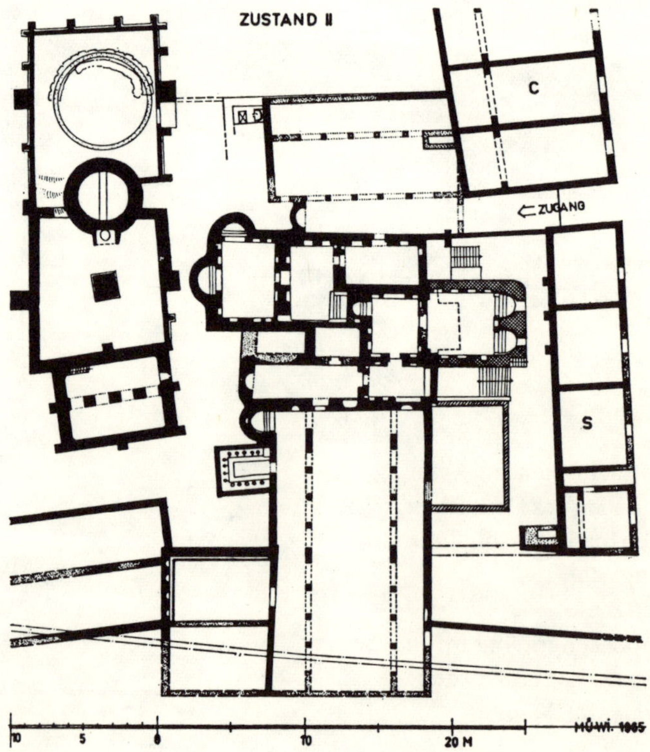

General Index